Tongues of Conscience

Robert Oppenheimer

Tongues of Conscience

War and the scientist's dilemma

R. W. Reid

Constable London

First published in 1969
by Constable & Co Ltd
10 Orange Street, London WC2
Copyright © 1969 by R. W. Reid
SBN 09 455890 6

Set in 11 on 12 pt Monotype Garamond
Printed in Great Britain
by The Anchor Press Ltd, Tiptree, Essex
and bound by Wm. Brendon & Son Ltd

There have always been in this world men of such keen intelligence that with their discourse they have been capable of infinite and various inventions that are as beneficial as they are simultaneously harmful to the human body. Wherefore, from day to day, we see unfolded before our eyes some of these that greatly astound our minds, and we remain so stupefied that when we think of them we are not able to control our faculties for some time, both from considering by what necessity or purpose these men were goaded as well as from contemplating the profound subtlety of their discoveries, which, in truth, are such that they are awarded the greatest commendation by noble minds.

Vannoccio Biringuccio: *Pirotechnia*
(*Director of Papal Munitions, 1538*)

My conscience hath a thousand several tongues,
And every tongue brings in a several tale,
And every tale condemns me for a villain.

William Shakespeare: *King Richard III*

Contents

Illustrations

Illustrations

Foreword

The problem of the conscience of the inventor, or the engineer, or the scientist, who finds that his work can be used either for the good or for the harm of mankind, is not a new one; gunpowder was first discovered by Taoist alchemists looking for elexirs of love. However, during the present century both the scale of the effect on society of the scientist's work and the way in which the scientist has acknowledged some of his responsibilities to society have changed dramatically. In this book I have chronicled the development of a number of scientific discoveries and inventions of the past hundred years which have been designed to take human life in war and which have influenced history, and I have tried to describe some of the attitudes and illuminate some of the motives of the men whose scientific abilities have so powerfully contributed to this influence.

Many scientists, by their willingness to discuss their work, have contributed to this book and I should like to acknowledge my debt to them. I particularly wish to thank my wife, Penelope, Mr Ronald W. Clark and Mr J. E. Stanfield for comments on the manuscript as a whole, and Mr Erik Bergengren, Professor H. A. Bethe, Professor Aage Bohr, Sir Lawrence Bragg, Dr H. A. Einstein, Professor O. R. Frisch, Professor S. A. Goudsmit, Dr Richard G. Hewlett, Dr J. H. Humphrey, Professor R. V. Jones, Sir Rudolf Peierls, Sir Michael Perrin, Dr V. Sidel and Professor R. R. Wilson for comments on some parts of the book.

R.W.R.

Tongues of Conscience

scientific life as, what might be called, a military chemist working on explosives, and in particular on propellants for artillery. For over 30 years he had been Chemist of the War Department, carrying out and publishing detailed research work from his official residence at Woolwich Arsenal. For his services to the nation he was knighted in 1891 and created baronet in 1893, the year before Nobel's Company's legal action.

James Dewar was 15 years younger than Abel, but his research covered a wider field both as a chemist and as a physicist. He was a great experimental scientist. He had carried out pioneering work on the liquefaction of gases and will be remembered for a long time to come by both scientists and picnickers for his invention of the Thermos flask. He had published papers on organic chemistry and had an interest in explosives. He was Fullerian Professor of Chemistry at the Royal Institution, and ten years after the legal action was knighted.

It was a so-called 'friendly suit', but Nobel, Abel and Dewar were all well aware that amity was in fact at stake. And this was not all. The costs in the case would be certain to be enormous and even these would be insignificant in comparison with the other financial implications of the verdict. But perhaps the most delicate feature at issue was the integrity of the two British chemists, both very pillars of the nation's scientific community. There were suggestions that the methods by which Abel and Dewar had come across the information from which they had patented their invention were, if not dishonest, then scarcely those expected of brotherly scientists and Victorian gentlemen. The roots of the events leading up to this investigation of character lay in some experimental work carried out 15 years before. Like so many scientific advances, it was work inspired by warfare.

The battlefields on which European nations had fought during the 19th century were characterised by a sight which had been seen on all battlefields where gunpowder had been used during the previous five centuries; after cannons, muskets and artillery had fired their first rounds the lines of troops would become hidden in a great pall of black gunpowder smoke. Unless sufficient wind rose to remove the shroud, it would stay and thicken for the rest of the battle, hiding men from both their own and their enemy's generals. During the 1870's several governments tried to find a solution to this problem and it was one of many to which Alfred

1: Scandal

On 29th January 1894 the case of *Nobel's Explosives Company Limited v. William Anderson* opened in the Chancery Division. The case at issue was apparently a simple and common enough one of infringement of an industrial company's patent. But it was not quite what it seemed. The names of the litigants in the action, a limited company and a private person, were mere substitutes for those of the real protagonists; these were, on the one side, a Swedish-born millionaire industrialist and, on the other, two distinguished British scientists; and not only were the rights of the patent at stake, so were the considerable reputations of the pair of eminent chemists.

The millionaire was Alfred Nobel, himself a most astute chemist and inventor. Thirty years before, he had patented his best known invention: dynamite. It had made the fortunes of the chemical explosives industry and of himself. The patent at issue in the High Court was one of his new inventions, the British patent of which had been acquired in 1892 by one of his own concerns in Scotland, Nobel's Explosives Company. It was this company which was taking legal action against William Anderson.

Dr William Anderson was the Director General of Ordnance Factories which supplied explosives to the British Army and Navy. It was one of his Royal Gunpowder Factory's products which stood accused of infringing Nobel's patent, and this product had been invented not by Anderson, whose role in the case was purely nominal, but by Professor Sir Frederick Abel and Professor James Dewar.

Frederick Augustus Abel, then aged 64, had spent his whole

Nobel applied his considerable inventive talents over a period of eight years from 1879. The result was *Nobel powder* or ballistite.

Ballistite was made up of approximately equal parts of nitroglycerine and nitrocellulose mixed with 10 per cent of camphor. In his British patent of 1888 Nobel described the nitrocellulose he used as being 'of the well-known soluble kind'. It was a casual phrase which showed the incomplete understanding of the cellulose molecule by chemists of that time; the choice of those words was to result in a great deal of bitterness for Nobel.

Ballistite's properties were quite remarkable. Besides being smokeless, or very nearly so, it was easy and safe to handle; it was more powerful than gunpowder; it left no deposit and the velocity with which it exploded was relatively low, making it particularly suitable for use in artillery, torpedoes and other armaments needing controllable propellants.

The significance of ballistite was for warfare alone, but such was this significance that Nobel was able to offer the patent for sale to European governments with no doubt of its ultimate commercial success. The Italian Government was more than interested. It introduced its armed forces to the powder and eventually bought the patent in 1889 for 500,000 lire. Meantime France's explosives monopoly, *l'administration des Poudres et Salpêtres,* had developed its own smokeless powder and was watching Nobel's activities with some alarm. Nobel, who had lived and worked in France for 18 years, now found himself accused unjustly of spying, his laboratory searched by police and his French factory prohibited from the manufacture of ballistite. Not surprisingly he soon moved his home and his laboratory to Italy to a climate more suited both to his delicate health and to his delicate commercial and political position.

Though France had failed him, Britain was another potential customer. In 1888 the British Government appointed an Explosives Committee to make recommendations to the War Office about technical improvements in modern military explosives. Ballistite was of course one of the outstanding innovations which would have to be considered and perhaps recommended for use in the armed forces. Acting as Chairman of this Committee was Sir Frederick Abel and, as members, his friend Professor James Dewar, and Dr Dupré, the Chemical Adviser to Her Majesty's Inspectors of Explosives. It seemed that the choice of Chairman could not have been more satisfactory for Nobel. He had known Abel intimately for many years; they had had meetings in London and Paris; they had corresponded on scientific subjects and had even had amicable disagreements which had ended only in increased mutual respect. Nobel could expect from Abel the informed judgment which many years' work in explosives research had given him.

When Abel and Dewar wrote to Nobel asking for details of his new discovery there was no reason whatever for him to have reservations: quite the contrary. Nobel had no hesitation in sending the complete specification of ballistite to the Committee and during the first few months of 1888 he carried out a friendly correspondence with Abel and Dewar, sent them samples of his material and allowed the Committee's Secretary to investigate his ballistic manufacturing process at Le Havre. Some of the letters from the chemists, chiefly those from Dewar, criticised several of the properties of the explosive powder: for example, the volatility of the camphor content gave it a variable ballistic quality. Nobel replied that he would experiment with a less volatile substitute for camphor.

Early in 1889 Dewar told Nobel that he and Abel were conducting a series of experiments looking for a substitute for camphor, and that they had obtained 'excellent results'. A few weeks later Abel was writing to Nobel to tell him that the excellent results had been patented. This was the first news that Nobel had which suggested that Abel and Dewar were turning from being collaborators into being potential competitors. The correspondence was continued for some weeks. At one point when it was suggested that Nobel's invention was being plagiarised Dewar replied, 'I hope Abel and I have some magnanimity left'. Nobel had reason to doubt. Before long the Royal Gunpowder Factory was beginning to manufacture a new smokeless powder: cordite, an explosive based on a patent by Sir Frederick Abel and Professor James Dewar and recommended to the War Office by the Committee on which they sat as being suitable for adoption as a propellant by Her Majesty's Army and Navy.

It was a most bitter blow for Nobel. All his commercial and inventive life he had been involved in patent disputes, but the source of this one took him by surprise and undoubtedly did little to help his already failing health.

Abel and Dewar's propellant was shown to consist of 58 per cent nitroglycerine, 37 per cent gun-cotton, and 5 per cent vaseline and acetone. The product could be forced through dies to form rods or cords: hence its name of cordite. The main feature by which it differed from ballistite was in the use of gun-cotton, which is an insoluble form of nitrocellulose and on which Abel was an undoubted authority, rather than Nobel's nitrocellulose 'of the well-known soluble kind'.

Eventually Nobel's Explosives Company, which owned the ballistite patent, decided to bring their action against Dr William Anderson as Director-General of Ordnance Factories. The Government had agreed to accept the court's ruling on Anderson as being binding for them.

The case, heard in Chancery in January of 1894, was both lengthy and highly technical. It had taken the staff of Nobel's Scottish laboratories almost three years to carry out experiments and prepare evidence for the hearing. Nevertheless, of most important scientific and legal interest is that the chemistry of nitrocellulose was by no means fully understood at the time and although its main practical properties were well known, it was not realised that the cellulose molecule occurs in a range of sizes and that the solubility of nitrocellulose depends on both its variable nitrogen content and its variable molecular weight. It soon became obvious as the case proceeded that a decision would hinge on the meaning of 'soluble' nitrocellulose, therefore the extent of the existing knowledge of the factors determining solubility was of great importance. In ballistite Nobel had used what he called nitrocellulose 'of the well-known soluble kind'. In cordite Abel and Dewar had used what they called 'gun-cotton, or the highest nitrated varieties of nitrocellulose which are not of the kinds known as soluble'. Nobel's Company's counsel, Fletcher Moulton, Q.C., argued his client's case with sweeping technical competence and some hearty wit. In his opening address he reviewed Nobel's considerable contributions to the world's knowledge of explosive. He summarised the work he had done on nitroglycerine, dynamite and blasting gelatine, leading to the ultimate refinement of the ballistite patent. It was, so Moulton told the court, the taming of nitrocellulose by nitroglycerine; indeed, it reminded him of 'a fervid Irishman speaking at the Union, who said it was universally known that starch, and particularly potato starch, caused an irritation which could only be allayed by the prompt application of Irish whisky'.[1]*

Many of the best known scientists of the day appeared to give evidence in court and made appropriate comments on the scientific integrity of Abel, Dewar or Nobel, depending on whether they favoured cordite or ballistite. Sir William Crookes, F.R.S., was called for the defence, and so was Sir Henry Roscoe, F.R.S., who named Abel as perhaps the highest authority on gun-cotton in the world. Sir Andrew Noble, F.R.S., formerly captain R.A., under cross-examination by Mr Moulton said there was no doubt of the indebtedness of the world to Mr Nobel for his inventions, 'if the world could be indebted for such things'.

Nobel, Abel and Dewar all gave evidence. Dewar, who has been described by one who knew him as 'of an impatient disposition and given to vigour of expression', soon found himself on the receiving end of Moulton's wit. Moulton brought up the subject of Nobel's letter referring to Dewar and Abel's remaining magnanimity. 'What magnanimity had you left?' asked Moulton, Q.C. 'I do not remember,' said the witness, 'but I know what I have left now.' 'Of the well-known soluble kind, I suppose,' retorted the learned counsel producing, as *The Times*[2] reported the following day, the effect of general laughter in court.

Moulton argued correctly that there is no hard and fast dividing line between soluble and insoluble nitrocellulose, but that there is a gradual variation of solubility between the two. But for all his technical clarity and wit, the result of the 13 days of the closely reasoned case was against his clients. Mr Justice Romer in his judgment said '. . . Mr Nobel in his specification in effect put aside the insoluble as being outside his invention . . .' And as for the British scientists, 'Sir Frederick Abel and Professor Dewar have not, in my judgment, been employing Mr Nobel's invention or merely colourably using the insoluble nitrocellulose as and for the soluble . . .'.[3] The action failed.

The Times, on the same day that it reported the conclusion of the case, developed Mr Justice Romer's judgment as a vindication of Abel and Dewar in a leader-page article. It reminded its readers how only a few months before there had been scandalous attacks on the methods of the Government's Explosive Committee and on the motives of the inventors of cordite. It rubbed salt in

* Notes and sources begin on p.334

Nobel's gaping wound by reminding him of the magnanimity of Abel and Dewar and how '. . . they were anxious to take him along with them in their researches'. The article ended with a smug snub of Nobel and a final sentence which the leader writer, if he lived long enough, might have had cause to regret: 'He has been fighting for the maintenance of a gigantic monopoly, and his failure liberates the trade from very heavy burdens. It also liberates the German Government from the obligation it formerly recognised to pay him royalties upon the cordite which is now coming into extensive use in the German army and navy.'

This was not the last word as far as Nobel was concerned. Although judgment was against him he knew that in some quarters there was sympathy. An appeal was lodged and was heard in July 1894 in the Court of Appeal. Mr Justice Romer's judgment was affirmed. Finally the case was taken to the House of Lords in February 1895 where again Nobel lost the decision. The total cost of the suit which the Company had to meet was £28,000.

At no time during the case was Nobel's integrity questioned; rather, the opposite was true and on several occasions it was assumed, and he was complimented. But in spite of the judgment of Mr Justice Romer and *The Times* concerning Abel and Dewar, there were still words of criticism heard against them. Mr Justice Kay in the Court of Appeal, while concurring in judgment of his colleagues, the Master of the Rolls (Lord Esher), and Lord Justice A. L. Smith, had some doubts about the ultimate decision. During the hearing of the Appeal he is reported to have used a number of metaphors far from flattering to Abel and Dewar: he reminded the court that a dwarf on the back of a giant can see further than the giant himself. He expressed sympathy for Nobel, considered his innovation to be 'a grand invention',[4] and pointed out that two clever chemists had got hold of the patent, had read it carefully, and with their own thorough knowledge of chemistry had used almost the same substances, and with a significant enough difference in one of them, had produced the same result. It was a harsh comment on the integrity of the two knights of the realm.

Nobel was bitter at the result. At one point he thought of writing to the War Office to ask for a guinea in recognition of his services to its explosives industry but was advised against it. He had lost a great deal of money, he had forfeited potentially vast sums which patent rights might have brought him, and he had lost faith in two distinguished scientists. He went off to write a play parodying the English legal system. The only consolation he could take was the knowledge that there were other heads in England besides Lord Justice Kay's which had shaken a little at the behaviour of Abel and Dewar. There was press comment and questions in the House of Commons. Both before the case and subsequently a number of negotiations involving the scientists were brought out into the open, some of which were considered questionable.

For example, it was revealed that while they were members of the Government's Explosives Committee, Abel and Dewar had accepted fees from Nobel's Explosives Company for acting as scientific advisers during an application by the Company for an extension of another of Nobel's patents for an explosive – blasting gelatine. More disturbing was the disclosure that Abel and Dewar had kept the foreign right of their patent and had sold it very profitably to foreign powers while the cordite specification was still being kept secret by the War Office.

The reactions of the public to the revelations of the cordite case are an enlightening comment on the image of the scientist in society during the closing years of the 19th century. The two chemists while sitting on the Government's Explosives Committee had carried out deals in explosives which were unmistakably to their own benefit, and in addition, though with the knowledge of the Government, they had sold weapon secrets to foreign powers. At this a section of the public was mildly scandalised, just as it would have been had any Englishman behaved in this unpatriotic fashion. But in spite of the judgment, and in spite of *The Times*, what appeared to be most reprehensible and which had been hinted at by Lord Justice Kay was that there had been shown to have been a breach of honour among the brotherhood of scientists. Two most distinguished men of their profession had loop-holed a patent of a third in a manner which fell only just within the law. The dividing line between this and actually stealing the invention was very fine. This was not the fashion in which Victorian England expected its most distinguished men of science to behave in private life.

The surprising aspect of the cordite scandal was that at no time during the whole case was a murmur raised that the making of military explosives for profit in times of peace was either unusual or not expected of a distinguished academic such as James Dewar. And even if it was expected that Abel, the military chemist, and Nobel, the tycoon chemist, should earn their living from war weapons, no suggestion was heard that science might expect finer ultimate motives from its workers. There was no expectation that if these scientists should be permitted this behaviour under certain circumstances, then they should be responsible to mankind for their products and should, because of their specialist knowledge as scientists, take positive action to see that their discoveries were restricted, or used for the good by attempting to educate mankind as to their work's benefits and dangers.

The only incisive criticism which *The Times'* leader article had had to make on the day following the judgment was that the result of the case prevented Nobel building up a gigantic monopoly in ballistite: and serve him right. It freed the armaments trade to make the explosive without any restriction. At the same time *The Times* took the trouble to remind the German Government that, as a result of the action in the English courts, it could now presumably stop paying Nobel his royalties, and that would be the end of the affair. This was the age of Victorian rationalism. If a scientist by his discoveries introduced to society something which would allow one man to kill another then it was society's job to work out how best to use it. These were still the dying years of the 19th century; they were years when scientists were expected to bear only the same responsibility as the next man. The conscience of the scientist, the creator, and his special duty to society were not yet anticipated. He did not have to concern himself with the consequences of his work.

2: The Prize

The scientist whose discoveries had affected the development of military explosives and that of warfare probably more than any other man since the age of gunpowder, the man who had emerged from the cordite scandal morally commended, yet full of bitterness, was the extraordinary Alfred Nobel.

It can scarcely be true to say that Nobel chose for himself the path his life's work was to take. His father and his brothers also earned their living as technologists, inventors and arms manufacturers and in his early years at least Alfred simply followed the family trade. But just as the attitudes of his later life were highly original and individualistic, so was his behaviour consciously extraordinary in his youth. He was a lonely child; by the age of 18 he was already deeply introspective and had begun an autobiographical poem. This is hardly surprising in a boy of 18, but the language in which he chose to write his poem was English rather than his native Swedish, and his confidence in his own intellectual ability was sweeping:

> ... When fellow-boys are playing,
> He joins them not, a pensive looker-on;
> And thus debarred the pleasures of his age,
> His mind keeps brooding over those to come.
> With an imagination made to scale
> The utmost heights to which the mind can soar. ...

> ... I have not shared the pleasures of the crowd
> Nor moved in Beauty's eye compassions' tear,
> But I have learned to study Nature's book
> And comprehend its pages and extract
> From their deep lore solace for my grief.[1]

Alfred Nobel was born in the northern outskirts of Stockholm on 21st October 1833. In the same year his father, Immanuel Nobel, filed a petition in bankruptcy after having failed to make a living as a housing contractor. Immanuel tried setting himself up as a chemical, industrial and arms manufacturer, but with such lack of success that he decided to leave his impoverished family behind and emigrate. Eventually he found his way to St Petersburg where his wife and children joined him in 1842. Things went better in Russia. The Russian Government became interested in 'the foreigner Nobel' and what was described in a communication to the Minister for War, as methods 'for destroying the enemy at a considerable distance'. These were the land and sea mines which Immanuel Nobel had only a few years before offered to the Swedish armed forces who had responded with conspicuous lack of interest and no money whatever. Russia provided the financial backing at a time particularly convenient for both herself and for Immanuel. During the 'forties he was able to build an engineering business, develop his inventions both connected and unconnected with armaments, and by 1854, which was the year of the outbreak of the Crimean War, be in a position to supply the Russian High Command with a variety of efficiently turned-out weapons. Nobel's products ranged from engines and guns for corvettes of the Russian Fleet to floating mines intended for the Anglo-French Fleet should it attempt to get at St Petersburg through the Baltic. Immanuel Nobel and his son Robert's efforts to protect Finnish waters were so successful that Admiral Napier reported that 'the Gulf of Finland is full of infernal machines'.

For a foreigner Immanuel Nobel had not done at all badly. Quite apart from the money he made from the Crimean War his services to Russia had been recognised even before it began by the award of an Imperial gold medal. The time had come in the life of the man who had spent about a quarter of his life in Russia, and now approaching 60, to begin to think of himself as having at least the foundations for a permanent stake in the country. But as the war ended so did the armaments contracts and Government promises. Immanuel, who had expanded his already large factory to meet anticipated armament needs, by 1859 was yet again bankrupt and starting back for Sweden with a good deal of bitterness and resentment towards Russia.

Inventors, engineers, armament manufacturers without national allegiances: this was the pattern to which the Nobel family had set by the time Alfred Nobel reached maturity. It was a way of life he was never to question and a discipline he was to accept and never attempt to change during the whole of his life. But Alfred Nobel was no conformist. Already by the time the Crimean War was turning up the family's first fortune he had the makings of an unusual and in many respects strange young man in whom the beginnings of paradox were evident. He had become a first-class chemist with an inventive turn of mind but yet had a passionate love of the English Romantic poets. He had already travelled about the world and had become an excellent linguist and yet he was a self-confessedly lonely young man. Physically he was weak, but mentally he was strong: the life he made for himself was to be physically cosseted but mentally unhappy.

It was during the family's more prosperous early period in Russia that Immanuel Nobel and the boy he called his 'good and industrious Alfred' began to look at the properties of nitroglycerine as a more effective explosive for tunnelling than gunpowder. The son's first significant invention was a method of detonating nitroglycerine with gunpowder. His patent of the *Nobel detonator* revolutionised the explosives industry; it was on this discovery that modern methods of blasting are based. But there was some dispute between the father and the good and industrious Alfred over the priority of the work. Alfred's astute business sense which sent him to patent dozens of his inventions was eventually to involve him in many wrangles over priority. That the first of these should start in his own family could only have given him some idea of the bitterness to come in the years ahead.

The father-son dispute smoothed itself out on the surface at least, but the next family tragedy which followed in a few months had harsher consequences. Immanuel Nobel had returned to Stockholm to try to set himself up yet again in the explosives business. In 1864 he rented from a local landowner part of an estate and a small workshop where laboratory tests could be carried out on the family's latest ideas. On 3rd September the workshop exploded, killing a mechanic, a young boy, a maid, a workman who was passing by, and Emil Nobel, Alfred's 20-year-old brother. It was a catastrophe for Immanuel. Within a month he had a stroke, as a result of which he was to be more or less bed-ridden for eight years.

At 31 this might have been the turning point in the career of Alfred Nobel, a young man who had enough diversity of scientific talent to move successfully into a less risky field. All that explosives and the armaments industry had given him so far had been a dead brother, a bankrupt and crippled father and a distraught (but adored) mother, along with pending lawsuits, an antagonistic Stockholm populace and a hostile police force. If ever the time was to arise when the physical, if not the moral, dangers of the consequences of his work should cause him to call a halt, then that time was now. But he did not call a halt. Nor, if it comes to that, did the father sink into a useless despair. Immanuel continued to turn out imaginative and apparently uncontrollable ideas from his sick bed, and Alfred went on to build an unrivalled industry.

In spite of the ban on the manufacture and storage of nitroglycerine in a populated area, which Swedish authorities imposed after the 1864 explosion, and in spite of the fear and suspicion which the blasting oil everywhere engendered, Alfred Nobel set out to rebuild the family business from precisely the same source as the origin of the disaster. Working on a barge anchored on Lake Mälaren outside the populated area of Stockholm he began to manufacture *Nobel's Patent Blasting Oil* at 2·50 Swedish crowns a pound. Already he had persuaded the Swedish State Railways of the superiority of nitroglycerine over black gunpowder and had had it officially adopted for tunnel blasting in Stockholm. Within a few weeks of the Stockholm explosion he was helping to found the world's first limited liability company in the explosives field and becoming its managing director, works' engineer and treasurer. Foreign companies were soon founded on the success of the Swedish operation. A factory at Krümmel near Hamburg was well situated to supply nitroglycerine not only to Germany but also to ship it to the United States, South America and Australia. Twice before 1870 it was destroyed by explosions.

At the age of 33 Nobel took a Greek word, *dynamis*, meaning power, for the source of the name for his newest invention. It was soon to become one of the best-known patented names in the world. Nobel had been trying to find a less dangerous substitute for nitroglycerine for some time. The oil was difficult to transport and, as a result of casual and – as Nobel frequently pointed out –

stupid methods of handling, had caused many deaths. In Britain, under the Nitroglycerine Act of 1869 the manufacture, use and transport of the substance was to be forbidden. Nobel's answer was to absorb nitroglycerine in kieselguhr, a fairly pure form of fossil silica, which gave a plastic explosive with an explosive power five times greater than gunpowder. This explosive, *dynamite,* was the basis of the future chemical explosives industry. It made Nobel and many others into very rich men.

Nobel spread his industry with great fixity of purpose into an empire. Already by 1873 he had established or helped set up factories in Scandinavia, Germany, Austria, America, Scotland, France, Spain, Switzerland, Italy, Portugal and Austro-Hungary. In 1876 the starting capital of the Hamburg company, which became known as Dynamit A/G, was 3½ million marks; it had risen to 5 million in 1888, 9 million in 1898 and 12 million in 1908. By the beginning of the First World War Dynamit A/G was the largest explosives manufacturer on the European continent, and by 1918, after having been a major source of high explosives throughout the war, its share capital was 36 million marks. Ranged against this armament output had been the factories which Nobel had created in Britain and which prospered at a rate similar to those in Germany. After the armistice the whole of the British explosives industry was amalgamated with Nobel's Explosives Co. as the nucleus. In the 1920's there were more mergers involving 17 British firms and a final amalgamation with other large chemical companies to become the Imperial Chemical Industries.

Nobel was long dead by the time the Great War had his factories attempting to out-produce one another from opposite sides of the holocaust, but similar situations most certainly arose during his lifetime. In 1868 he went into partnership with a Frenchman named Paul Barbe, a man Nobel was later to describe as being the owner of an india-rubber conscience. Barbe, on the partnership's behalf, tried to interest the French Finance Ministry in dynamite but the French explosives monopoly pressurised the Ministry into refusing to become involved. However, when the Franco-Prussian war broke out in 1870 and the French found their defences being blown up by the Nobel dynamite they had pecuniarily rejected, new attitudes were brought to bear on the Ministry decision. A French Government loan financed the building of Nobel's factory in the South of France; it was set in production at frantic speed so that French troops were being supplied with dynamite before the end of the war.

One of Nobel's official biographers maintains that Nobel loathed war and that as soon as the Franco-Prussian armistice was signed he rushed to put dynamite 'on to the right market for which it was intended'.[2] It is undoubtedly true that Nobel, even then as a relatively young man, disliked the consequences of war; but in the case of the Franco-Prussian conflict he, or his business partners, had little alternative other than to try to re-market dynamite. Once the war had stopped and military consumption of dynamite dropped, it had then to be directed into a highly competitive market at mining and tunnelling. There is no evidence to show that Nobel had any finer motive other than an acute business sense for making the best use of his factories.

Nobel was fully aware of the great benefits which mankind could accrue from his inventions and some of these were spectacular even during his lifetime. The blasting of the Alpine passes and tunnels of the 19th century were magnificent engineering achievements which could not have been done without dynamite and its derivatives. The St Gotthard, Simplon and Arlberg tunnels all exist as monuments to his scientific invention. Mankind's acknowledgments of his contributions to its welfare gave Alfred Nobel considerable satisfaction, as they would most men. In 1868, along with his father, he had been awarded the Letterstedt Prize of the Swedish Academy of Sciences 'for outstanding original work in the realm of art, literature or science, or for important discoveries of practical value to mankind'. The joint prize had been awarded to Immanuel for his work on nitroglycerine and to Alfred for his invention of dynamite. But many years later when his mother died Nobel wrote to her executor asking for only a few keepsakes, though showing his deep resentment of his father having accepted the credit for the discoveries he believed to be his own. He wrote: 'The Letterstedt Medal may as well come to me. I understand perfectly what my mother meant when she wrote, "It belongs to Alfred Nobel". My mother knew a lot of things that the public is not aware of.'[3]

This letter was written, not by a young man still in need of public recognition, but by a man nearer 60 than 50 who was a public figure and a millionaire.

Nobel was plainly a very lonely and self-pitying man to whom life brought not much happiness. He is reported as having led an unindulgent sort of life, neither smoking nor drinking, playing neither cards nor party games. He suffered from a number of illnesses and in letters criticised his doctors in the sarcastic and acid fashion which characterised his writings. Once when he was asked by a brother for autobiographical details he wrote: 'Alfred Nobel – his miserable existence should have been terminated at birth by a humane doctor as he drew his first howling breath. Principal virtues: keeping his nails clean and never being a burden to anyone. Principal faults: that he has no family, is bad tempered and has a poor digestion. One and only wish: not to be buried alive. Greatest sin: that he does not worship Mammon. Important events in his life: none.'[4]

It is a sad, revealing letter. Nobel makes no attempt to hide his self-pity and realises that his loneliness springs in part from having no family of his own. This he calls a fault, but at the same time he fails to appreciate that what he considers to be one of his greatest virtues, his wish never to be a burden to anyone, must similarly contribute to his loneliness. His wish not to be buried alive is no flippancy and his will had a directive to have his veins opened and death confirmed before cremation.

In 1876 Nobel was living and working in Paris, was highly successful and was at the height of his inventive powers. He had never had a secretary and, possibly in an attempt to ease his loneliness and his secretarial problems in one fell swoop, tried the unorthodox solution of putting an advertisement in a Viennese newspaper; it read: 'A wealthy, educated, elderly gentleman living in Paris is looking for a mature lady with a knowledge of languages to act as his secretary-housekeeper.'[5] The result was a meeting between Nobel and a member of the Austrian aristocracy seeking a diverting job, Countess Bertha Kinsky. It seems that Nobel was not particularly disturbed that Bertha's maturity was that of only 33 years, and she even confided some pleasure to her memoirs when she found that, far from being an elderly gentleman, he was only 43, and sound in wind and limb: not handsome, but not ugly.

A *rapport* sprang up between the pair: he showed her his poems and she told him of her unhappy, unfinished love affair in Austria. Even though Bertha's stay with Nobel was brief, it was long enough for him to confide in her the beginnings of the rationalisation of his position as a scientist-armaments inventor. She was sufficiently impressed by his reasoning to record it in her memoirs:

> Alfred Nobel could devote only one or two hours a day to me, for his work bound him. Once again he had a new invention in mind. He said to me, 'I would like to produce a substance or a machine of such frightful, enormous, devastating effect, that wars would become altogether impossible.' A week after my arrival Herr Nobel had to make a trip to Sweden for a short time, where he was building a dynamite factory; the King himself had sent for him.[6]

Bertha did not stay long enough to learn more of either his proposed methods of promoting peace or of his influential connections. While Nobel was in Sweden she had a telegram from her lover which swept her back to Vienna and to his arms in a fashion which even the 1870's must have found conspicuously romantic. But though the secretarial-housekeeper attachment came to nothing this was by no means the last she and Nobel were to see of each other. They did not meet again for several years, but during this time the thoughts and writings of both became more concerned with war and what their individual attitudes to it should be.

Nobel wrote to a Belgian pacifist how his dream for the future was 'little different from Diocletian's cabbage patch, watered by the waters of Lethe. The more I hear the guns roar, the more I see the blood flow, plundering legalised, and the revolver sanctioned, the more vivid and intense this dream of mine becomes.'[7] Nobel was beginning to see how this dream, this 'rose-red peace in this explosive world', should be achieved: it was by the technical perfection of weapons of war, and it was Bertha's memoirs which faithfully recorded the crystallisation of his attitudes.

When they met after a separation of over ten years, Baroness von Suttner, as Bertha had become, had considerably broadened her experience of life. Part of her happy marriage had been spent in Caucasia during the Russo-Turkish wars where she had seen and been deeply affected by the suffering around her. She became influential in international peace movements and the renewal of the old but brief acquaintance with Nobel was an attempt to gain some support of her cause. He was not entirely convinced by all her arguments, but there is no doubt that she managed to exert her influence on him. And her charm did not go unnoticed. In

C

1890 after the publication of her novel *Lay Down Your Arms* he wrote to congratulate her. The letter was in French and ended, 'You should not cry *"à bas les armes"* since you use arms yourself – the charm of your style and the greatness of your ideas – these will carry further than the Lebels, the Nordenfelts, the de Banges, and all the other infernal implements.' The letter ended in English, 'Yours for ever and more than ever, A. Nobel'.[8]

Nobel was convinced that in relation to other armament manufacturers his position as an inventor was quite singular; it was, moreover, honourable because his motives were different from theirs. Even before Bertha von Suttner had published her great plea for peace he had written from Paris in reply to a letter in which she had invited him to visit her and her husband, 'At the moment all the "Dynamiters" of the world (the dynamiters are the directors and administrators of the dynamite companies) have gathered here to tease me with their affairs: conventions, projects, deceptions and etc. and I ardently wish that a new Mephisto would come to enrich Hell with these evildoers.'[9] They were evil, but he, the inventor of dynamite, was not one of them.

Between Bertha and Nobel there existed a mutually admiring respect of aims but an amicable mistrust of methods. When Bertha asked for financial support for her peace propaganda Nobel sent £80 in sterling and a letter saying that what she really needed was not money but a programme for peace. Resolutions, congresses, speeches and banquets might be her methods of convincing nations of the necessity for peace, but he left her in no doubt that they were not his. And disarmament: 'To ask for disarmament is practically to make oneself ridiculous.' He felt that for a peace movement to have any success it should begin in a more modest way; it should recommend the practice of England where in matters of doubtful legislation the country can introduce an act with a provisional validity of two years. For instance, could not European governments undertake for a period of a year to submit all differences between them to a special tribunal? Or if they were not prepared to do this could they not postpone hostile action until the expiration of a stipulated period and passions had subsided?

In 1892 Bertha took a leading part in the Peace Congress at Berne. Nobel visited her there briefly and invited her with her husband to spend some time with him in Zürich. Nobel had

already seen the commercial possibilities of aluminium metal and
had in Zürich a large aluminium yacht which he used to take his
visitors on trips round the lake. For Bertha these were days of
idyll. The old *rapport* was strong. 'We spoke about a thousand
things between heaven and earth.' They even agreed to write a
book together in which they would attack everything of misery
and stupidity which the world had preserved. Animatedly they
talked as the shining boat made its tours of the lakeshore for her
to admire the villas of the Swiss silk manufacturers.

> 'Yes the silkworms have spun all that,' said Nobel.
> 'Perhaps dynamite factories are even more profitable than silk
> mills and a little less innocent,' she remarked,

perhaps at the risk of seeming a shade waspish. But he had his
defences ready.

> 'Perhaps my factories will put an end to war even sooner than
> your congresses: on the day that two army corps can mutually
> annihilate each other in a second, all civilised nations will surely
> recoil with horror and disband their troops.'[10]

Bertha faithfully recorded the details of their little sorties in her
memoirs. If Nobel had not had a mission for his science and
industry before, then he had one now. Perhaps the ultimate
weapon could make war impossible more quickly and more surely
than any other method yet known. He was prepared, however, to
investigate any other avenues which might lead towards peace,
and being a rich businessman he decided to delegate the investigation to a paid assistant. The man he chose was a one-time Turkish
Ambassador to Washington named Aristarchi Bey, now impoverished and seeking a little patronage. Aristarchi's exact duties were
never clearly defined, but were roughly to keep Nobel informed
about pacifist activities and to act as the source of useful ideas.
After a year the relationship came to an end, with Aristarchi, like
so many of Nobel's other acquaintances, squabbling over money.
Nobel had had to conclude that his expensive diplomat had not
'converted to the cause sponsored by Mme de Suttner a single
person of any importance'.

According to the cynical autobiographical extract which he
wrote for his brother, Nobel believed one of his principal virtues
to be that of 'keeping his nails clean'. In other words, he was a man

of honour. But it seems never to have occurred to him that the standards by which he led his life were peculiarly paradoxical. In public he led a life of morality, and in private one of enterprise. If the enterprise should offend public morality then it should be abandoned. But clean nails in private did not necessarily mean the same thing as clean nails in public. For example, he once thought of getting a controlling interest in a Swedish newspaper and asked his nephew to make a few quiet inquiries for him. His nephew misinterpreted the motive and had to suffer an avuncular reproach:

> You seem to imagine that my object is to influence the market, but a newspaper owned by me would rather tend to arouse opposition. It is one of my peculiarities that I never consider my private interests. My policy as a newspaper owner would be to use my influence against armaments and such mediaeval survivals, but to urge that if they are to be manufactured they should be manufactured at home; for if there is one branch of industry which should not be dependent in any way upon imports from abroad, it is surely the armaments industry.[11]

In other words his policy as a newspaper owner would be quite different from the policy which had helped to turn him into a millionaire.

His dealings with Bertha von Suttner were stamped with the same contradictory attitudes. The period of their most intimate acquaintance came only a short time after he put ballistite, an explosive which had little application outside warfare, on to the market. During the same period that he wrote so passionately to Bertha approving her 'eloquent pleading against that horror of horrors – war' he was involved in his correspondence with Abel and Dewar telling them how he would try to improve ballistite in the hope that the War Office would find it useful.

It was not only in his attitude to the stock market and to armaments that Nobel maintained his public and private morality, and it was Bertha who almost found him out. For many years he kept an affair going with a young lower-middle class Austrian girl he had found on a business trip, and whom he maintained in some style in the cities and spas of Europe. Her name was Sofie Hess. She was pretty, petulant, demanding and uneducated. He was indulgent, in need of affection, jealous and a fatherly, rich Henry Higgins. In return for her willingness to allow him to try to turn her into a lady he pandered to most of her whims, and one of these was to be addressed as Madame Nobel. When Bertha one day discovered from a florist that flowers had been addressed to a Madame Nobel, she wrote to congratulate him on his marriage. In his letter of reply Nobel flirted with both Bertha and perhaps also with the truth. He explained away the flowers by suggesting that the woman for whom they were intended 'was without doubt my sister-in-law'. Unwittingly Bertha left Nobel open to posterity's tut-tutting by printing his ambiguous letter in her memoirs as an example of his charm.

Sofie, unlike Bertha, but like many of Nobel's acquaintances, turned out to be a source of disillusionment for his later years. She had a child by a Hungarian cavalry officer and blackmailed Nobel (and the executors of his will) into keeping her in comfort for the rest of her life. The 216 letters from Alfred to his 'sweet Sofiechen' were sold to the executors and their existence not disclosed by the Nobel Institute until 54 years after his death.

The consolation of his later years was his inventive ability which never left him. During his lifetime he had been responsible for over 350 patents, both connected and unconnected with explosives, and yet in spite of or, as he might have argued, because of, his deep concern for peace he still continued experiments on new weapons: silent guns, expanding linings, shell fuses, propulsion charges, rocket projectiles. He had, of course, no financial worries which drove him to work in this field. After the cordite affair his vast business enterprises, far from slumping, developed well and increased his already enormous income. Now he claimed to his co-workers that the technical challenge which weapon development offered him was so formidable that he found it irresistible. The financial rewards were as nothing: but yet he marketed his products. The temptation which technical achievement offered to Nobel in the 1890's was to be repeated to other scientists in the century which followed. Fifty years after he in part saw his motivation, physicists were still being intrigued by the sweetness of invention and they were claiming the possibility of, but still seeking, the ultimate weapon which Nobel believed would make war unthinkable.

Nobel died of cerebral haemorrhage at his villa in San Remo, Italy, on 10th December 1896. During the last years of his life he had had to suffer the almost indecent irony of having doctors prescribe him nitroglycerine which he had to take internally. He found that doctors called it trinitrin to avoid rousing any fear in a patient who might link its chemical name with its explosive properties.

He died as he feared he would: without family or friends nearby to ease the last hours. He died a multi-millionaire: unmarried and without an heir. It was to Bertha von Suttner, three years before his death, that he first confided his wish that the fortune he had gathered should be used to encourage man to peaceful habits. He wrote:

> I should like to allot part of my fortune to the formation of a prize fund to be distributed every five years (let's say six times, for if we have failed at the end of thirty years to reform the present system we shall inevitably revert to barbarism). This prize would be awarded to the man or the woman who had taken the greatest step towards the idea of general peace in Europe.[12]

He went on to make quite clear to Bertha that the disarmament which she saw as the answer to peace was not his idea of a solution, and that disarmament could only be achieved slowly. He also went on to advocate a peace-keeping force to be formed by states acting together which would take action against any aggressor. The principles he proposed were very similar to those eventually adopted by the United Nations.

Nobel's last will was made public in January 1897. It listed some private bequests and then directed that the remaining capital should be invested and that the interest should be distributed as prizes to those

> . . . who, during the preceding year, shall have conferred the greatest benefit on mankind. The said interest shall be divided into five equal parts, which shall be apportioned as follows: One part to the person who shall have made the most important discovery or invention within the field of physics; one part to the person who shall have made the most important chemical discovery or improvement; one part to the person who shall have made the most important discovery within the domain of physiology or medicine; one part to the person who shall have produced in the field of literature the most outstanding work of an idealistic tendency; and one part to the person who shall have done the most or the best work to promote fraternity between nations for the abolishment or reduction of standing armies and the holding and promotion of peace congresses. . . .[13]

There was something of Grand Guignol in the eventual execution of the will; it was plagued with difficulties of every kind. In the first place, although Nobel's last will was explicit, it was not strictly legally valid. He had drafted it alone and without any legal advice. It was at first contested by a number of Nobel's relatives who had been virtually disinherited. Then there were the accounts of such unsuspected dependents as Sofie Hess to consider: Sofie was able to wave in front of the noses of the executors letters addressed in Nobel's own hand to his 'sweet Sofiechen'. Even the Swedish people were said to be piqued; Nobel had suggested that it should be the Norwegian parliament which should award the great peace prize.

When at last the necessary international legal arrangements had been made to give validity to the will it was possible to make a final settlement of 31,000,000 kronen to the Nobel Foundation. The Nobel Prizes, all ready and waiting to be awarded to the most distinguished work in science, literature and peace, were conceived just in time to coincide with the arrival of the 20th century.

For scientists, at least, the Nobel Prize established itself, as soon as the century got under way, as the most coveted statement of recognition of achievement which could possibly be gained, and it has continued to hold the same position in the league table of prizes ever since. Men who believe themselves to be in the running for the prize and whose daily lives are normally uncompetitive still wait nervously, as men did 60 years ago, for the telegram carrying the good news telling them they have been chosen to join the band of specialist immortals. One recent prizewinner filled a bath with champagne so that his friends could share with him the good news and the prize money.

The money which Nobel used to institute his prizes was the direct result of the combination of his own indisputable scientific achievements and his facility as a businessman. The fact that during his life he was in the van of scientific thought and invention gave him a great deal in common with the prizewinners who later benefited from his business sense. One not unusual common factor was that the end product of the technological achievement could be used in war and to take human life. Nobel's pre-eminence in the explosives field meant that a not inconsiderable proportion of Nobel prize money is derived from armaments manufacture. The pre-eminence of Nobel prizewinners in their scientific fields means that in times of war they may find themselves heavily involved in the production of weapons and facing similar and, in some cases, the identical moral problems which Nobel had to face. The history of the changing character of the moral conscience of the 20th-century scientist can be traced in the attitudes of Nobel prizewinners.

Speaking at a Nobel anniversary dinner shortly after the first atomic bombs had been dropped, Albert Einstein, himself a Nobel prizewinner, remarked that physicists in 1945 found themselves in a position not unlike that of Nobel:

> Alfred Nobel invented an explosive more powerful than any then known—an exceedingly effective means of destruction. To atone for this 'accomplishment' and to relieve his conscience he instituted his awards for the promotion of peace. Today, the physicists who participated in producing the most formidable weapon of all time are harassed by a similar feeling of responsibility, not to say guilt.[14]

Nobel was not the wicked inventor whose only wish was to use war and its horrors to thicken the lining of his pockets, as some writers would have him be; nor was he the incorruptible personality whose great inventions for peace were twisted into weapons by political power, as one official biographer considers. He behaved as he felt the times demanded.

He lived in a great rationalistic age. If, as Einstein suggests, he had a troubled conscience then he had allayed it long before the institution of his Peace Prize by reasoning that the deadlier the weapon, the more effective it will be in wiping out war. He gave no evidence whatever of suffering from a feeling of guilt. As far as he and his age were concerned the limit of his responsibilities was reached with each piece of scientific creativity. He would invent, manufacture, distribute and make money. If mankind chose to wage war with his creation, then it was mankind's responsibility and not his. As the 20th century was born, this was still the attitude expected of the scientist. Both attitude and expectation were to alter as the century was to grow to middle age.

3: The Young Genius

The zenith of Alfred Nobel's inventive powers coincided with the rise to power of one of Europe's most considerable statesmen, Otto von Bismarck, and both men had signal, if unrelated, interests in the outcome and duration of the Franco-Prussian war.

On 18th January 1871 at about midday 60 detachments from Prussian regiments moved into the Bourbon Palace at Versailles. The occasion was the proclamation of the German Empire. On a beflagged platform at one end of the Hall of Mirrors stood the Prussian King William. Even the king was apparently somewhat overcome by the grandeur of the moment: an American visitor noticed that his legs trembled slightly. Directly in front of the platform stood the man who had raised the king to the heights of German Emperor: the Junker, Bismarck. It is unlikely that Bismarck's legs, beneath the great pair of cuirassier's boots, were trembling. He was a man at the peak of his confidence and of his political career, having built a state of a kind entirely different from any that had previously existed. Hanover, parts of Denmark, and Poland, Alsace-Lorraine and Bavaria had been moved into the Empire; Prussia was now the predominant power in Germany and the Second Reich had begun.

Bismarck's successes put the intellectually progressive groups in Germany in a difficult position. He had unified Germany and given her power and stability: this was admirable. What was not admirable was the manner in which he achieved his ends. He was to continue advocating his own successful methods: 'Place in the hands of the King of Prussia the strongest possible military power, then he will be able to carry out the policy you wish; this

25

policy cannot succeed through speeches, and festivals and songs, it can only be carried out through blood and iron.' Blood and iron! Jewish intellectuals in particular found the predicament a difficult one to resolve. How far dare they allow themselves to become influenced by the Prussian character which was now dominating Germany? And even if they wanted it, would they be permitted to assimilate with the ruling class? Were Bismarck's remarks on the subject at Versailles perhaps encouraging? He had said, when discussing whether his own Junker breed could not be made more tractable through marriages with Jews, that where there were examples of such marriages the results were 'not at all bad'. He was not too happy with the thought of Jewish men marrying into Junker families, and thought it to be preferred if one 'brought together a Christian stallion of German breed with a Jewish mare. There is no such thing as an evil race, and I don't know what advice I might one day give my sons.'[1]

There is no doubt that in this country of unsure racial tolerance, where power and force were glorified, the Jewish father had many more problems than Bismarck in giving advice to his son. If Hermann Einstein, like most others of his Jewish middle-class, was an admirer of Bismarck, then either he did not succeed in passing on this admiration to his son, or he tempered the admiration with so much caution that his son made his own very different appraisal.

Albert Einstein was born at Ulm on 14th March 1879 and a year later the family moved to Munich. It was a family which was typical of many: it considered that its Jewish customs had been assimilated by the environment and by the Christian way of life in the west German city. The boy was allowed to grow up as an agnostic and was taught nothing of Jewish traditions, habits and food laws. Every Thursday the parents invited a poor Russian Jewish student to their midday meal, which might have been a continuation of an old Sabbath custom, or it might have been nothing more than an example of Hermann Einstein's warmheartedness.

The young Einstein was not a child prodigy. He was a good mathematician, but otherwise an average performer at the Catholic elementary school to which he was sent. Although the family considered it had been assimilated into the Christian Germany of the 1880's the boy knew well enough of a difference between himself and his classmates: he was a Jew; they were not. There was no prejudice shown towards the boy, but he was sensitive of the difference: his classmates called him Honest John. He was also attuned to other strata in Munich life. He saw the similarity between the methods of the teachers at the elementary school and the methods of discipline of the Prussian army, and he liked neither. At the age of 10, he entered the Luitpold Gymnasium in Munich and compared the teaching methods at his old and new schools: 'The teachers in the elementary school appeared to me like sergeants, and the gymnasium teachers like lieutenants.' The child had sensed the class structure of the Prussian army in the German way of life and already had learnt which behaviour to expect from which rank.

The next few years were to form a very deep and lasting impression of his own German nation on the boy. In the streets of Munich he could watch the military parades and the stiff lines of obedient men marching with some happiness to the beat of drums. And in the citizens of Germany he could see the adoration of this military mentality and the worship of the Prussian methods of force. He did not like what he saw either in the soldiers or in the citizens. During this period of his childhood there were noises-off in the German political scene which just a few Jews heard as faint warning shots. On one occasion a small group of Pan-Germans demonstrated both inside and outside the Austrian Parliament. They were violent and anti-Semitic. Forty years later their temperament and methods were to be taken over by the Nazis.

Einstein's father's electrochemical business ran into difficulties when the boy was still at school. He closed down his factory and decided to migrate to Italy and rebuild the business in Milan. The boy was left behind in Munich to study at the Gymnasium. To be taught in a city where, in addition to the discipline, even the methods of teaching were mechanical and military was one thing: to be abandoned there at the age of fifteen was another. Either because of a real or an imaginary nervous breakdown he managed to persuade a doctor to give him a certificate recommending that he be allowed to join his family in Italy for six months. When he reached the gentler arms of his parents he immediately asked to renounce his German citizenship.

This total rejection of the country of his birth might well have been the action of a child sickening for the protective warmth of his family, but it was an attitude in which Einstein was to be remarkably consistent for the rest of his life. Einstein's colleagues of his later years all recall him as a man of great gentleness and kindness. His life was one which did not breed stories of envy and gossip. Yet there is some doubt that the portrait of an invariably angelic human being of infinite tolerance towards his fellow creatures, which every biographer attempts to paint, bears any resemblance to the young Einstein. He was an emotional, sometimes rough, young man. Einstein's son, Hans Albert, is able to recall the occasions when his father's consistency in the dislike of physical methods of discipline slipped occasionally and he received a well-deserved clout behind the ear. This roughness showed itself in ways other than that of the minor sin of an angered parent: it was both a significant physical and mental quality. Einstein was to devote his life to a search for a grand design of truth, but it was to be an aggressive search. In science he was to organise a revolution, but in bringing it about he was to show the typical characteristics of a revolutionary. He could ride heavily over his opponents; if he felt that his critics were attempting merely to smooth out insignificant wrinkles on his grand design, then he could be tactless, rude and hurtful. His reactions to pedantry, or stupidity, or cruelty were those of uncompromising intolerance. In particular his attitude to Germany and all things German, which formed itself so early in his life, was one of committed suspicion and condemnation. Even in his later years he refused either to forgive or to forget the war crimes of Germany as a nation and he never attempted to compromise by indicting Naziism rather than the German nation. This sensitivity to the touch of the heavy hand of Germany, this rejection of what undesirable shapes of mind it might attempt to mould, were characteristics which already forcefully influenced his attitudes by the time he was 15. They were the roots from which grew some of the most significant actions of his life.

In spite of the fact that his parents were entirely irreligious Jews, Einstein went through the customary youthful phase of a religious fervour. It did not last long. Even as a boy in Munich he had begun to read popular scientific books which had him reasoning that the stories of the Bible might not be true. The result was that he rebelled against the idea of being forced to attend religious services as much as he rebelled against the idea of having to submit to Prussian school military discipline. And so by the time he reached Milan he was not only ready to renounce his German citizenship he was also ready to break his new links with religion.

He had become what he called a freethinker. He taught himself the principles of mathematics and calculus and by the time he was 17 had enrolled as a student in mathematics and physics at the Swiss Institute of Technology at Zürich under such great mathematicians as Hurwitz and Minkowski. And by the time he had completed his course at Zürich he should have been ready to apply his freethinking to the world of science. He had renounced his religiosity, he had renounced his nationality, and he had attached himself to a new country. He was ready to go. But neither the breaks with the old nor the attachments with the new were as final as he believed. Jobs of any kind were not easy to come by in this first year of the new century, and he had to earn his first meagre living as a tutor. When at last he found full-time employment, at the surprisingly good salary of 3,000 francs a year, Einstein considered himself very fortunate. He had been interviewed by the director of the Swiss Patent Office in Berne, a man who apparently recognised the independence of Einstein's thought and the value this would have in patent assessments, in spite of the lack of any appropriate training.

Landing this job was an important event in Einstein's life and in his career. Firstly it made him financially independent for the first time in his life, and therefore able to marry Mileva Maritsch, a physics student with him in Zürich; secondly it gave him a great deal of free time to apply to theoretical physics. Exactly 20 years after Einstein had been working at the Berne patent office, Leo Szilard, a young Hungarian Jewish physicist who had just completed his doctorate, found himself in Berlin unable to get a job or to make a living. Einstein by this time was famous and successful, and since the two often discussed physics together Szilard asked Einstein what he should do in such a depressing position. Einstein replied by telling him about his job in Berne and suggested that Szilard should apply for a job in the Berlin Patent Office. Einstein went on, 'You know, this was not a great mental strain. It was not very hard work for me. And I had the great advantage

29

that I was under no moral obligation to lay golden eggs. This was the happiest time of my life.'[2] Szilard, however, did not take Einstein's advice, though he was to seek it again later on a more crucial occasion.

The fact that the patent office held Einstein under no moral obligation to 'lay golden eggs' on its behalf gave him great freedom both in his thought and in his work. It was here in Berne that he produced golden eggs for himself and for physics. As a result of the work he did mainly in the evenings in his small apartment he was able to publish in the space of a few months of 1905 five papers. Some of the ideas in these papers were to rock the existing foundations of science and help build them afresh. One of the papers dealt with 'Brownian motion' and showed how the atomic nature of matter could be verified; another was a contribution to quantum theory, which was at the time the most recent revolutionary development in physics. Either of these two contributions to physics would have assured Einstein a place in scientific history books. It was the third paper, however, on which the greatest part of his fame was to rest for the rest of his life. This was called 'On the Electrodynamics of Moving Bodies'. It was the foundation of the special theory of relativity.

Einstein's talent was soon recognised, even though perhaps not quickly enough for the likes of a young scientist with confidence in the validity of his new radical theories. Within a few years the university portals which had seemed so firmly bolted on their insides cautiously opened.

Until this stage in his career Einstein had had little enough reason to worry over disadvantages arising from his racial origins. After a university post in Zürich he was offered a professorship in theoretical physics at the German University in Prague. There a condition of employment was that a teacher should belong to a recognised church. Einstein was obliged to enter his religion as 'Mosaic', as the Jewish religion was called in Austria at the time. Under the old Austrian Empire Prague University had been divided into two parts, Czech and German, in an attempt to settle political quarrels between these two nationalities. The German population in the city was small, however, and about half Jewish. For much of its cultural life, therefore, it looked to its Jewish element. So, to the bitterly independent Czechs, the ways of the

Jews were the ways of the German nation, whilst to some Germans the Jews were, as always, the Jews. For the first time in his life Einstein now had close-hand experience of the dilemma of a small Jewish community.

As in other European cities a Jew had humiliating difficulties in finding employment. Only in government offices were jobs made available, and even these required some influence and graft in high places. Franz Kafka, for example, considered himself lucky to find work in a semi-government insurance office. In Prague at the time there was a group which included Kafka, Hugo Bergmann and Max Brod, which had been driven to becoming interested in Zionism and the possibility of establishing an independent enclave of Jewish thought. They failed to interest Einstein. He had seen the situation of the Jew and had disliked it. But it did not worry him: he still considered it unimportant.

In 1912 Einstein returned to Zürich, this time as professor. But although in this, the country of his adoption, the temperament of the people was more to his liking than in the country of his birth, there still remained a powerful attraction in Germany. A great deal of art can flourish in isolation, but not science; since a scientist requires the stimulation and encouragement of other scientific minds, scientific Mecca have always flourished and Germany in the early years of the 20th century was such a Mecca. Berlin was a strong magnet to Einstein, particularly since the Kaiser was enthusiastically finding finance for important research institutes in the city. The offer of the directorship of the still unbuilt Physics Institute of the Kaiser Wilhelm Gesellschaft, together with a professorship at Berlin University with no distracting administrative duties and no obligatory lectures but with a comfortable salary, was a very great temptation for Einstein. The offer was made by the distinguished scientists Max Planck and Walter Nernst, who took the trouble to travel to Switzerland to try to regain their brain drain. They took with them an additional carrot to offer: membership of the Royal Prussian Academy. This was one of the highest accolades of German science.

Einstein accepted. The job and the intellectual atmosphere it offered clearly appealed to him. Membership of the Prussian Academy was a great honour which he also accepted readily enough: he had no cause to offend Prussian science, nor Prussian scientists such as Max Planck, whom he liked and admired. However, he undoubtedly succumbed to this part of the temptation to move to Berlin with his tongue in his cheek. Throughout his life he was always ready, sometimes irrationally, sometimes jokingly, but never without provocation, to attribute racial characteristics to the Prussians. Many years after Einstein's entry into the Prussian Academy Patrick Gordon Walker, then an Oxford don, told him of a story he had heard being spread in Germany by Nazis: that Einstein had stolen papers, including the relativity theory, from the body of a dead Prussian officer. Einstein replied, 'No *Prussian* could possibly have worked out anything as intelligent as relativity theory.'[3] It was a bad joke in reply to a bad joke, but his awareness of the race was never anything but serious: 'These cool blond people make me feel uneasy; they have no psychological comprehension of others. Everything must be explained to them very explicitly.'[4] In the spring of 1914 Einstein left Switzerland for Berlin.

When war broke out the anti-Allied propaganda in Germany was as naïve and unreasoning as that of the anti-German propaganda in Britain and France. The intellectuals in Germany were as susceptible to the nationalistic fervour of the times as any other section of the community and it was not long before musicians, painters and scientists were making statements which in their terms were as aggressive as the actions of the military. The result of this hubbub of patriotic sentiment was a document which in later years was to acquire considerable notoriety: *The Manifesto to the Civilised World*. This paper was published in October 1914 and was signed by 93 of the most distinguished and apparently enlightened men of German culture. Among it's signatures were 15 natural scientists including Haber, Nernst, Planck and Röntgen: three musicians, including Humperdinck, the theatrical producer Max Reinhardt – as well as historians, critics and poets. Many who signed the Manifesto later claimed that they authorised their signatures on the strength of a telegram: considering its contents this would seem the only acceptable explanation why so many thinking men should have subscribed to such an ill-considered document. In addition to the expressions of patriotism common in wartime writings, such as the disclaiming of Germany from war guilt, the denial of German atrocities, and the accusations of Allied atrocities, it included such sentences as:

> ... But in the East the ground is soaked with the blood of women and children slain by Russian hordes, and in the West the breasts of our soldiers are lacerated with dum-dum bullets. No one has less right to pretend to be defending European civilisation than those who are the allies of Russians and Serbians, and are not ashamed to incite Mongols and negroes to fight against white men.

In its final paragraph it said:

> . . . Without German militarism German civilisation would be wiped off the face of the earth . . . believe us, believe that to the last we will fight as a civilised nation, to whom the legacy of a Goethe, a Beethoven, and a Kant is no less sacred than hearth and home. . . .[5]

For Einstein this pairing off of the incompatible values of German militarism and German culture could not have been more unfortunate and he did not sign. Within a few days he was approached with a counter petition from the heart specialist Georg Nicolai: *A Manifesto to Europeans*. It called for a collective peace in Europe and by implication was highly critical of the 93 intellectuals. But when Nicolai circulated his petition in the University of Berlin he could find only three men to add their signatures to his, and one of them was Einstein.

This gesture by the signatories required some courage, made as it was in the Berlin of 1914. Nicolai, as a result of his activities, lost his professorship and was eventually interned in Graudenz fortress. Einstein, however, was protected from accusations of traitorous behaviour since he had kept his Swiss citizenship. There was not and never had been a vestige of German patriotism in Einstein and, from his friends at least, he did not attempt to hide the fact. To his close friends he even went so far as to hope for an Allied victory so that the power of Prussia would be finally crushed.

He continued to live and work in the peace and quiet of Berlin. He disliked the war but was lucky to be able to turn his back on it. In the city there were no sounds of gunfire, no real fear of seeing enemy troops in the streets, and no involvement. Einstein was able to continue to work at the theories he believed had no relation to warfare whatever, and whose consequences could not directly affect either soldiers or civilians. There was no reason for him to suspect that within a quarter of a century the nature of the work he had already performed would force him to commit himself to help initiate momentous decisions for the enhancement of war.

Though Einstein was to have a quiet war without involvement, many other scientists were to become deeply committed. Members of Britain's most august scientific bodies offered what services they could to the national cause. Several months after the war had begun, for example, the Managers of the Royal Institution held a special meeting to consider their most recent letter from the Comptroller of Munitions Inventions and reminded him that 'the Managers are most anxious to render every possible assistance to His Majesty's Government in relation to the present War'.[1] James Dewar of the Royal Institution was soon to become consultant to the Air Board on the liquefaction and storage of oxygen; Rutherford and W. H. Bragg were to carry out anti-submarine work for the Admiralty Board. The spectrum of war work to which scientific invention could be applied, almost overnight, was discovered to be vast, though in the main it was the older men who had been left behind to make these applications.

As a group, men with specialist scientific and technological skills were not kept back from the Front either by Germany or by the Allies, though in a few cases they were channelled into departments of the armed forces where it was thought they might be most useful. Bragg's son, for example, W. L. Bragg, found himself in France with the rank of Second Lieutenant. There he became technical head of Sound Ranging and using microphones placed close behind the front-line to locate enemy guns. It was while he was struggling with this sound-wave technique of measuring the positions of the German batteries that the 25-year-old Bragg heard that he had been awarded the Nobel Prize. His father told him by letter that they were to share it for their work on the analysis of crystal structure using X-rays. The curé who owned the house in which Bragg and his colleagues were living brought up a bottle of wine from the cellar to celebrate the event.

Working with Bragg in France were some of Britain's brightest young scientists: E. N. da C. Andrade, C. G. Darwin, A. S. Russell and others who after the war could be expected to help shape the future of British science. Another group of scientists was kept in Britain at Farnborough to carry out research at the Royal Aircraft Factory. It was housed in what became known as the Chudleigh Mess. Forty years after this group had left Farnborough one of its members counted up the distinctions his colleagues had managed to collect in later life. These included three peerages, five knighthoods, three Nobel Prizes (F. W. Aston, G. P. Thomson, E. D. Adrian), and a cluster of Professorships and Fellowships of the Royal Society.

It was the same story in Germany. Several of Einstein's close scientific colleagues took commissions in the army: Walter Nernst and Fritz Haber both became Captains and worked on the development of poison gases. Haber had Jewish origins but was nevertheless considerably influenced by the Prussian view of military power. He was well pleased and even flattered by his rank, though some of his colleagues considered the 'spiritual leader' among chemists to be under-valued. When the British officer, General Hartley, who had been deeply involved in Allied gas warfare, visited the Rhineland after the war, a German industrialist was confounded when Hartley explained that he had become a soldier by leaving his lectureship in Oxford for his war-time commission. 'How is that possible?' he was asked; 'You a General, and Haber only a Captain!'[2]

It was Haber who was given the credit for having introduced chemistry in the service of Germany's war. It was he who developed the technique which in 1915 blew chlorine over the French and Canadian lines. He was given command of the Prussian War Ministry's organisation for gas warfare, gas protection and gas education to which he recruited chemists as private soldiers, N.C.O.'s and officers. His was not a back-room job. He appeared on the front-line to watch his techniques in operation, where one of his colleagues observed his 'cold-bloodedness, courage and disdain of death'.[3]

One spring day in 1915 he left the trenches for a short visit to his home: that day his wife, who reputedly desperately resented his total involvement in the gas war, committed suicide. On the same evening Haber travelled out to the Eastern Front where he knew he was expected.

But Haber's love affair with the military, its manners and its hierarchy was atypical. For most scientists in the field, British or German, rank had no great significance; what was important, besides the outcome of the war, was how they should resolve the great cleft in their loyalties. They had grown up in an age when the international character of science had been strong. Many of Britain's young scientists, such as those in France with W. L. Bragg, not only knew and respected their German counterparts but had lived and studied in Germany, spoke the language well and had taken their doctorates in Berlin or Vienna. How could they now be expected to lend themselves to the cause of embittered nationalism?

At a wartime meeting of the Royal Prussian Academy in Berlin the great chemist and Nobel prizewinner Emil Fischer stood up to warn-off his fellow academicians from extreme jingoistic attitudes with the words: 'Whether you like it or not, gentlemen, science is and always will be international.' But outside scientific circles no concessions were made to the brotherhood of science, and those who wished could pile the coals of propaganda on to the fire of hatred. When the British Association met in Manchester in 1915, its President, Sir Arthur Schuster, was attacked by a section of the press because he had a German name. During the same meeting Schuster was given the news that his son had been wounded in the Dardanelles.

Within scientific circles there was a strange half-truce. In spite of the fact that Britain and Germany were hard at war, letters still managed somehow to pass between scientists of each country giving news of the progress of research, asking after colleagues and looking forward to future cooperation. There was a firm belief that the interests of science transcended the present holocaust and that it was mutually understood that scientists could commit themselves to the demands of war only in the present temporary and regrettable circumstances.

So in 1915 Meyer in Vienna was writing a hearty letter to Rutherford in Manchester via a United States consulate. And in the same year Rutherford was writing to Geiger, a brilliant young physicist serving in the German Army. But as the war wore on the subject matter of letters became more poignant and less chatty. There began to appear warnings of losses to physics which might be irreplaceable. Geiger's reply to Rutherford came through a neutral country:

> ... I was on the Front up to middle of October but got ill and had to spend nearly ten weeks in bed. It was a rather bad rheumatism with fever and my legs were very much swollen and stiff. But now I have it over and am doing service again, but am not at the Front yet. I expect I shall be in the field again in about a fortnight.
> I think nearly everyone of my colleagues whom you know is in the field. Dr Rümelin (also Reinganum and Glatzel) fell in one of the first months of the war. W. H. Schmidt is also dead; but from what I have heard he must have been in a very bad and hopeless condition all last year and so it was perhaps the best for him. I sometimes hear from Professor Hahn who seems to stand the dangers of war very well and is apparently quite happy in his military position. Dr Schrader is in the field artillery and is quite well up. . . .[4]

And of course it was not only the soldiers of the front-line of his country's enemy who wrote to Rutherford; he had similar letters from his own protégés. One of the brightest stars of British physics before the war was H. G. J. Moseley. Moseley's pedigree was outstanding. His father was a Fellow of the Royal Society as had been both his paternal and maternal grandfathers. His education had been what might be called classically scientific for his day: King's Scholar at Eton and Open Scholar in Natural Science at Trinity College, Oxford. During the four years between Oxford and the war, Moseley had convinced all the physicists he had worked under, and particularly Rutherford, that there was a brilliant scientific career stretching in front of him. One of his most significant pieces of work had been the crucial discovery that the properties of an element are determined by its atomic number – the number of units of positive charge carried by the element's atomic nucleus; he also devised a method involving X-rays of determining atomic numbers.

Moseley was at a British Association meeting in Australia when war was declared. As soon as he returned to England he joined the Royal Engineers and became a signalling officer. In April 1915 he was writing to Rutherford, full of youthful anticipation, and saying that he hoped his Brigade would be in France before the end of the month. In July of 1915 a friend was writing to Rutherford pointing out that Moseley might be of more use away from the Front working on some scientific problem. Immediately Rutherford tried to use his influence to get Moseley transferred, but too late. Moseley's Brigade left for the Dardanelles on 13th June 1915. On 10th August Moseley was involved in the severe fighting at Suvla Bay, and caught by a Turkish bullet from a flanking attack only 200 yards away from where he was trying to telephone an order. A fellow officer wrote to Moseley's parents, with perhaps the bluntness of youth:

> Let it suffice to say that your son died the death of a hero, sticking to his post to the last. He was shot clean through the head, and death must have been instantaneous. In him the brigade has lost a remarkably capable signalling officer and a good friend. . . .[5]

Later it was learnt that Moseley, in his will made on active service, had bequeathed his apparatus and his considerable wealth to the Royal Society.

Later that year Rutherford wrote to *Nature*:

> It is a national tragedy that our military organisation at the start was so inelastic as to be unable, with a few exceptions, to utilise the offers of services of our scientific men except as combatants in the firing line. Our regret for the untimely death of Moseley is all the more poignant because we recognise that his services would have been far more useful to his country in one of the numerous fields of scientific inquiry rendered necessary by the war than by the exposure to the chances of a Turkish bullet.[6]

This chance bullet, perhaps more than anything else, brought home to the British the shortsightedness of their use of scientific manpower. And the lesson of Moseley's death was still being quoted by German scientists in 1945.

Germany in 1915 could also offer a biting example of how chance and science in a critical combination had affected the fortunes of the nation. But for Germany it was perhaps the most critical piece of good chance of the whole war. It could be argued with considerable justification that had it not been for Fritz Haber's method of ammonia synthesis and Wilhelm Ostwald's method of conversion of ammonia to nitric acid, Germany could not have continued the war beyond the end of 1916. Ever since men such as Alfred Nobel had shown that nitroglycerine, nitrocellulose and several other nitrated organic compounds could be used to manufacture easily controlled and powerful explosives, nitric acid, which is the essential nitrating agent, had become a vital commodity for any country hoping to wage a war. In 1914 the chief source of the nitrate salts from which nitric acid could be manufactured was Chile, so that the control of the sea-lanes to and from South America became one of the most vital of the early prizes to be fought over. It was the British Navy, oversure of its rule of the waves, which lost the first of the major sea battles for the precious nitrate. On 1st November 1914 off Coronel on the coast of Chile a British cruiser squadron was routed by a squadron of Germany's new and inexperienced Navy and chased off to the Falkland Islands. The First Lord of the Admiralty, Mr Winston Churchill, took under his control the organisation of the revenge for the débâcle. Within six weeks the British fleet emerged from its hideout refitted and reinforced with more ships and took an uncompromising toll of Graf von Spee's squadron. Only one German cruiser escaped from the second encounter and even she was subsequently hunted and scuttled: Britain was left master of the South American shipping routes and controller of Europe's source of nitric acid.

The battle of the Falkland Islands could have been the emasculation of Germany. Not only were nitrates imperative for the explosives industry, they were also vital as sources of fertilisers. With no other large nitric acid source available it should have been only a matter of months before Germany could no longer fight, and only a few more months beyond that that the country could no longer eat. But the answer to the Navy's shortcomings came from Germany's chemists. In 1909 Haber had filed a patent which described how ammonia could be synthesised from nitrogen and hydrogen at high temperatures and pressures, and two years before that Ostwald had filed a patent showing how ammonia could be burnt in air over a platinum catalyst to form nitric acid. It was mere experimental chance which dictated that these critical discoveries should be made a few years before the war rather than either one should come a few years after it. Shortly before the outbreak of hostilities Karl Bosch took over the development of Haber's process and turned it into a large-scale industrial operation. And so, in spite of the Falkland Islands disaster, Germany's chemists had given the nation the choice of fighting on, if necessary for another four years. This saving chemical grace sharply emphasised the high value Germany could place on her scientists.

Had Germany won the battle of the Falklands and wrested the control of the sources of nitrate from Britain there would have been a different and sadder tale to tell. Britain's capability of synthesising ammonia on an industrial scale in 1914 was scarcely enviable. On 15th May 1916, the day before the Managers of the Royal Institution's special meeting to consider the urgent enquiry of the Comptroller of Munitions Inventions, Professor Sir James Dewar had written in some *Notes for the Managers of the Royal Institution*:

> One German Company (Badische) between the years 1909 and 1913 took out no less than thirty-three Patents in this country on Synthetic Ammonia alone, that are all available for the Government Departments to use. This splendid development of Professor Haber's Invention was effected by the combined working of the huge Chemical and Engineering Staff of the Badische working in cooperative and at great expenditure during some five years. What hope can be entertained that the Royal Institution laboratory with a Staff ignorant of this kind of Work, even with the aid of a sympathetic Professor, could add any new knowledge that has not already been published?[7]

Meantime, however, the interest James Dewar had shown many years before in the manufacture of explosives, combined with the inventive skill of a Jewish scientist, made chemistry a handsome contributor to the British war machine. Cordite was now desperately needed in large quantities as a shell propellant, and the acetone which was one of its vital ingredients was being obtained by a process of bacterial fermentation of cornstarch, discovered by the organic chemist, Chaim Weizmann, in 1911. It was Weizmann's war work on acetone which brought him into contact with leading British politicians, so enabling him subsequently to argue persuasively for a Jewish national home in Palestine.

Of any distinguished French scientists who might have considered withdrawing from actions or decisions involving them in the war being fought over French soil, Marie Curie was the one from whom such behaviour would have been most easily expected and understood. By birth she was Polish, having come to the

41

Sorbonne as a student in 1891. She had married a Frenchman, the physicist Pierre Curie, who at the height of their combined creative careers had been killed in violent, strange circumstances: apparently by walking into the path of a horse-drawn six-ton wagon in the Rue Dauphine. By nature she was an introvert and her husband's death as was to be expected had driven her to the solitude of her laboratory bench where she worked long hours, almost wilfully neglecting her health. It would have seemed ordinary and predictable if in 1914 Marie Curie had packed her equipment and moved as far from the front-line as possible, quietly to continue her work.

But ordinariness was not one of her characteristics: she was a persistent, uncompromising and brilliant scientist who, like her husband, had a highly attuned and unusual conscience. With Henri Becquerel they had shared the Nobel Prize for 1903 for their work in radioactivity. On 6th June 1905, Pierre Curie had spoken before the Swedish Academy of Sciences in the Nobel lecture on behalf of himself and his wife. His subject was the element which together they had discovered: radium – a substance which promised remarkable cures for cancer and which had opened new pathways of discovery for science. The Curies were well aware that the creations of modern science can be used by the unscrupulous against the common good; they were also aware that the man who had created the prize they were being given had been troubled by the dilemma which this situation presented. Said Pierre Curie:

> One can imagine that in criminal hands radium might become very dangerous, and here one may ask oneself if humanity has anything to gain by learning the secrets of nature, if it is mature enough to profit by them, or if this knowledge is harmful. The example of Nobel's discoveries is characteristic: powerful explosives have permitted men to perform admirable work. They are also a terrible means of destruction in the hands of the criminals who lead the peoples towards war. I am among those who think, with Nobel, that humanity will derive more good than evil from the new discoveries.[8]

It was Pierre Curie who induced a radioactive burn on his own arm by deliberately exposing it to radium, and so drew the attention of the world to the dangers of radioactivity. Forty years later its effects were still being observed after the first atomic bomb.

Persistently Marie and Pierre Curie went to great extremes to avoid reaping any financial benefit from their work. They made no attempts to patent their methods of separation of polonium and radium and when American industrialists asked for details of their process of radium purification they unhesitatingly gave away whatever information they considered might be helpful. The motive which inspired the Curies to act in this selfless manner – because to act otherwise 'would be contrary to the scientific spirit' – was admirable; but the consequences to the Curies' own work could have been disastrous, and the question arose whether by refusing to profit from their work they actually hindered the progress of science. Only a few months before his death in 1906 Pierre Curie was writing to a rich woman pleading for money for a laboratory because, in spite of his eminence as a Nobel prize-winner, he was getting insufficient funds from the State to be able to continue his work with the facilities he needed. In fact a great deal of the most rewarding work he and his wife had done together on radium and polonium had been carried out in primitive conditions in an old shed.

Marie Curie was well aware of the criticisms of indulgent righteousness which were levelled; in an autobiographical note written later in life she wrote:

> A large number of my friends affirm, not without valid reasons, that if Pierre Curie and I had guaranteed our rights, we should have acquired the financial means necessary to the creation of a satisfactory radium institute, without encountering the obstacles which were a handicap to both of us and which are still a handicap for me. Nevertheless, I am still convinced that we were right.
> Humanity certainly needs practical men, who get the most out of their work, and, without forgetting the general good, safeguard their own interests. But humanity also needs dreamers, for whom the disinterested development of an enterprise is so captivating that it becomes impossible for them to devote their care to their own material profit.
> Without the slightest doubt, these dreamers do not deserve wealth, because they do not desire it. Even so, a well-organised society should assure to such workers the efficient means of accomplishing their task, in a life freed from material care and freely consecrated to research.[9]

It would have seemed reasonable to assume that Marie Curie classed herself with the dreamers and not with the practical men.

Yet when war came she was one of the first to remember what she believed to be the general good, and to act in a thoroughly practical fashion. The French Government was attempting to float war loans for which it made an appeal for gold. Without hesitation she decided to contribute her second Nobel Prize (awarded for chemisty in 1911), which she had left untouched in Stockholm in Swedish crowns, and took in her Nobel medals to be melted down. The bank official could not bring himself to accept these small symbols of her great achievements; she, on the other hand, had no sympathy for his idolatry.

Shortly afterwards she began to organise radiological units so that the war wounded could benefit from the new X-ray techniques which had been developed. In addition to installing and training personnel for 200 radiological rooms she equipped 20 motor cars, one of which she drove herself at the Front. During the time that she was director of this service over a million men were treated at her posts.

There was a strange codicil to Marie Curie's war work. In common with other civilians who had distinguished themselves during the fighting, she was offered the cross of the Legion of Honour in recognition of her work. She refused the offer. However, her daughter, Eve Curie, maintains that her mother, had she been proposed for chevalier of the order *as a soldier,* would have accepted. It was as though she wished her role as a civilian scientist hived off from her role at the Front. Just as scientific dreamers ought not to be concerned with industrial enterprise, so should scientists and their research not be concerned with war: science, according to her fast outdating ideals, was pure and should at all costs not be contaminated with the responsibilities of its applications; these could well be left to others to sort out.

Einstein said of Marie Curie: 'Of all celebrated beings she is the only one whom fame has not corrupted.' Einstein's own attitude to the war was Marie Curie's case taken to the extreme: he had as yet not committed himself in any way and took neither a direct nor an indirect part in the hostilities. But in spite of these extreme examples, this war had produced the beginnings of an important change in attitudes in others. Critical scientific good fortune, as was Germany's in having the Haber ammonia synthesis in 1914, and bitter tragedy, such as was Britain's in the death of Moseley, had taught their lesson. It was now no longer possible to divorce science from its applications any more than it was possible to divorce science from war. The most important role which a scientist could play in wartime was as a scientist. Rutherford had acknowledged the fact in Moseley's obituary. What was now emerging was that the ubiquity of scientific invention and the esoteric nature of scientific thought would be capable of putting scientists in a singularly responsible position in society. It was also emerging that scientists could no longer expect to throw the responsibilities for the applications of their work on to politicians or on to soldiers since, such was the shape their work was taking, soon only they would be in a position fully to understand its significance and its practical consequences for mankind. The life Marie Curie had commended for dreamers – one 'freed from material care and freely consecrated to research' – was not always to be either possible or desirable.

During the Great War there were of course examples of men who were at the opposite extreme from Einstein's attitude of non-involvement and who committed both themselves and their science fully. Allied and German chemists both carried out work on poison gases and saw it applied. Haber and Nernst were named in the Treaty of Versailles as war criminals to be handed over by Germany for trial before an international court. No serious attempt was made to put this part of the Treaty into practice.

Only a few months after the war Haber was given the Nobel Prize for chemistry in recognition of his pre-war discoveries. But the ceremony was soured. The odium of poison gas was by now firmly stuck to his name: objections to the award were raised by French scholars. Haber, the gas manufacturer, was considered 'morally unfit' for the prize paid for by Nobel, the dynamite maker. The odium was to remain for many years, and even Rutherford found an excuse not to meet Haber so that he would not have to shake hands with the chemist who had earned himself the title of 'inventor of chemical warfare'. The title was no more deserved than the odium. Haber's methods were merely ingenious developments of techniques which have a long history in applied science, and many chemists on both sides of the fighting line could equally have shared the disapproval of society which Haber almost alone had to carry.

His fervent patriotism continued to flower even after the war;

he spent some time trying unsuccessfully to isolate gold from sea-water in order to enable Germany to pay off her war reparations. But Fritz Haber's contribution as a scientist to his country's economy had already been of glittering worth; the petty pride he took in his promotion to Captain was a small self-indulgence when set alongside the enormity of the value of his ammonia process.

Now, however, the question could happily be posed, would it ever again be necessary for scientists to be faced with the dilemma of having to choose whether, and if so how, to apply their work to the kind of total war they had just experienced? Some, scientists included, had no doubts: the most valuable consequence of that war was the end, for good and all, if not to warfare as such, then to the scale of bloodshed, destruction and the debasement of human values of 1914–18. On 24th November 1918, two weeks after the armistice agreement had been signed, Niels Bohr wrote from Denmark to Rutherford in England:

> . . . we feel an immense relief now the war is finished. All here are convinced that there can never more be a war in Europe of such dimensions; all the people have learnt so much from this dreadful lesson, and even here in these small Scandinavian countries, where for good reasons, there certainly was not much aggressive military spirit before the war, people have got to look quite differently than before at the military side of life.[10]

For Einstein the cloistered, academic and somewhat rarefied atmosphere of Berlin had many advantages and within a year of his arrival he was ready to publish what was to be his greatest work – the general theory of relativity. It is interesting to compare how much the anticipated value of the application of scientists' work has changed since 1915. During the First World War the work of many scientists, and in particular of theoretical physicists such as Einstein, was considered to be of so little significance that there were few restrictions on scientific publications. Already by March of 1917 Einstein's work had been so widely publicised that the Astronomer Royal of England was pointing out its significance and suggesting when it might be verified by experiment. In March 1919 a solar eclipse was expected when measurements could be made on how much the rays of light from a star are bent in passing through the Sun's gravitational field. Even though wartime conditions might have made it impossible to carry out the experiment, in London the Royal Society and the Royal Astronomical Society still went ahead with preliminary plans to observe the eclipse in South America and in West Africa. It was a remarkable situation that at the height of a terrible war, scientists of one of the protagonist nations should be preparing to verify the work of a scientist of the chief enemy.

It was another example of the strength of the international brotherhood of science and yet a powerful comment on the detachment which many scientists still felt towards the realities of the world. This detachment was to persist still for several years.

But as far as the Royal Societies' plans to verify Einstein's work went there were happy enough consequences. When the armistice was declared Sir Arthur Eddington gave details of the expeditions being organised to observe the eclipse; less than a year later, on 6th November 1919, at a dramatic Royal Society meeting, looked on with some severity by the portrait of Isaac Newton, he announced the verification of Einstein's general theory.

The Times gave a full report of the meeting on the same day that it remembered 'the glorious dead' of a year past. This expedition by a group of British scientists to verify the predictions of a German scientist in the same year that the two nations had been attempting to bleed each other to death did nothing but good for international relations at a time when amity was most needed. There was another important consequence: Newton had been superseded and a new star had risen: Einstein.

Einstein's genius among scientists had been both recognised and acknowledged for 14 years but it was only now that the public at large became aware of him. It is difficult to analyse the sources from which the world-wide fame of Einstein sprang so suddenly. The public was undoubtedly well aware that there were strange happenings in the traditionally unshakeable temples of science; in Britain, for example, some of its most honoured scientists had had to acknowledge that the stature of its greatest scientific genius, Newton, had been bent in places by a German-Swiss clerk. Moreover the popular press found copy in Einstein and in the apparently far-reaching consequences of his theories. Newspaper photographers discovered a highly photogenic and, for a time at least, tolerant subject: his was a face of character: drooping, kindly eyes and wrinkles of humour surrounded by a leonine mane of hair. The habits of the man were a little irregular; already some of the characteristics expected of the absent-minded professor were beginning to show: he lived a simple life un-cluttered by possessions and any of the outward trappings of success; when there was no need to be careful he was careless about his dress: sometimes he wore no socks.

All these qualities, combined with the publicised qualities of the man, kindliness, gentleness and warmth, would still not have been sufficient to turn Einstein into the international figure he was to become. The missing ingredient in this recipe for public fame was the apparently incomprehensible nature of Einstein's work. For a few years after the publication of the general theory of relativity only a limited number of scientists familiarised themselves with it in detail. Its abstruse nature became legend and absurd stories sprang up around its esoteric significance. It was even rumoured that there were few men in the world who were capable of understanding the theory. One story had it that a newspaper reporter had approached Sir Arthur Eddington and said that he had heard that there were only three people who were truly able to understand Einstein's work. 'Really?' was supposed to have been Eddington's reply. 'And who's the third?' Popularisations of relativity theory appeared in the newspapers and magazines of a world which, after four years of war, was delighted to read something other than stories of trenches, wounded, rehabilitation or peace conferences. In most cases the popularisers failed to remind their readers that if the author of relativity theory had been best able to express his work in non-mathematical language then he would probably have done so. The satisfactory outcome of this great burst of popularisation was that a part of physics, in the name of Einstein and in the word 'relativity', entered common culture. The tousle-haired man became the subject of cartoons, the butt of jokes ('Tell me Dr Einstein, what time does this station stop at the next train?'); and because of his singular casual Bohemian appearance he became the epitome of the scientist. The unsatisfactory outcome of it all was that Einstein was assumed to have a deeper insight than other men into subjects of which he claimed no special knowledge. Vaguely it was known that his work had revolutionised scientists' concepts of space and time, and therefore it was believed that in some way Einstein was dabbling with space and time and perhaps even dabbling with things quite near to God. Whatever the nature of the reasons the result was surprising: Einstein was the first scientist to become a world figure in his lifetime.

His views and judgment were sought on any aspect of life which a man of great wisdom might be expected to hold views and give judgment. In spite of his refusal to approve the workings of the German war machine his stature in Berlin academic circles had risen high. During the revolutionary disturbances in the winter of 1918 in Berlin he was asked to mediate in the students' revolt. The Rector of the University and part of his staff had been imprisoned in the Reichstag by students, and Einstein, his opinion respected by both sides, was called on to negotiate their release.

E

From Russia Einstein had invitations to inspect the new régime: he refused, partly because he feared that an open gesture of friendliness might automatically brand him as a Communist sympathiser. From London came an invitation to describe the results of his work in *The Times*: he accepted, and pulled the editor's leg for having tried to dissociate him and his work from Germany and things German. From America there came the friendliest intimations of waiting hospitality; behind his acceptance of these invitations was a deeply considered political motive, but there was also the more mundane reason that the offer of a free trip abroad was not in those days so easily come by for a scientist. And finally from France came an invitation to attend the Collège de France, the nation's most distinguished scientific academy. Hatred between France and Germany had diminished only a little since the war and the risk of some hostility being shown to this member of the Prussian Academy was very real. Einstein's visit to the Collège passed without incident, but he had to refuse an invitation to a session of the French Academy knowing that many members had threatened a walkout if he walked in.

When Einstein arrived in Paris for the visit he was met by newspaper reporters who asked a favourite question of the times; an American physicist named Miller had proposed an experiment to disprove the theory of relativity: what would Einstein's reaction be if Miller were proved correct? Einstein replied, 'I would be astonished.' The answer was amusing enough but not what the reporters wanted to hear and so they pressed for a more comprehensive reply; they asked what personal consequences Einstein would suffer. His answer was the parable he often told when the occasion suited; it was the manner in which he had replied to the editor of *The Times*: 'That's very simple,' he said. 'These days the Germans consider me to be a German, but the French consider me to be a Jew. But if Miller proves to be right and I am wrong, the Germans are going to consider me a Jew and the French are going to consider me a German.'

This was not the sour joke of a bitter man, it was the prophecy of an astute scientist. Einstein was amused by the prejudice against his scientific philosophy and saw the irony of the prejudice against his nationality; what now worried him was that there were increasing signs that the prejudice was shifting to a racial basis.

These remarks were made eleven years before Hitler came to power.

He had already realised how easily he could abuse his position of privilege and when the Berlin correspondent of *The Observer* requested an interview in 1921 he refused: 'My views and opinions are currently being given far too much weight, and I have, therefore, grown reluctant to speak in public about subjects on which I possess no special competence. Further there can be no doubt that there exists, here in Germany, considerable irritation with my pacifist and general political orientation.' However, there were two subjects on which he showed no reluctance to speak out in public. The first, in spite of the irritation in Germany, was pacifism, and the second, in spite of his non-religious upbringing, was Zionism. Though these were the two causes which he never hesitated to support, in one case he was to suspend his conviction at a crucial time.

His pacifism and dislike of the military mentality stemmed from his childhood and his schooling in München:

> This topic brings me to that worst outcrop of the herd nature, the military system, which I abhor. That a man can take pleasure in marching in formation to the strains of a band is enough to make me despise him. He has only been given his big brain by mistake; a backbone was all he needed. This plague-spot of civilisation ought to be abolished with all possible speed. Heroism by order, senseless violence and all the pestilent nonsense that goes by the name of patriotism – how I hate them![1]

On this subject of pacifism he was never anything other than outspoken. To a group of visiting Americans he said: 'My pacifism is an instinctive feeling, a feeling that possesses me because the murder of men is disgusting.' A reporter for *The New York Evening Post* succeeded where a few weeks earlier his colleague on *The Observer* had failed; not only did he manage to interview Einstein, he was sent away with the directive, 'Do not omit to state that I am a convinced pacifist, that I believe the world has had enough of war.'

The Zionism which Einstein supported was to some extent thrust upon him as a burden needing a recognisable porter, but it was a cause for which he felt an increasingly great need. Spiritually he was not a Jew. His agnostic family had made no attempt to teach him Jewish customs which might make him feel a part of the race; it probably even considered itself assimilated into Christian Germany. But Einstein's experiences, both as a child and as a young scientist struggling for recognition in Switzerland, Czechoslovakia and Germany, taught him that whereas he might attach no particular significance to his racial origins, they were all too important to others. He had been successful, for the time being, in not suffering personally from the prevailing prejudice. His colleagues, however, had not all been so successful: and there were signs that the prejudice might become more harmful. Even in 1921 he told his colleague Philipp Frank that he did not expect to be able to stay in Germany for more than ten years: it was an amazingly accurate forecast.

There were political aspects of Zionism also which did not appeal to Einstein. Of necessity, if it were to succeed it would have to create a strong sense of nationalism and this was a sentiment which he considered to have been the bane of Europe and the world. But he had seen the sufferings of Jewish people at first hand and he clearly foresaw that given the wrong circumstances these sufferings might take on new proportions. And so when Chaim Weizmann, the Manchester Professor of Chemistry, now the world's leading Zionist, had invited him to travel on his American tour of 1921 to raise funds for a Hebrew University in Palestine, Einstein had accepted.

The idea of the establishment of a Jewish national home was not universally popular. Einstein's alignment with Weizmann's cause met with disapproval from German Jewish intellectuals who believed that again the time was ripe for assimilation. But in the United States Einstein had been an unqualified success. There he had cashed-in on the credit his world fame as a scientist had given him and allowed Weizmann to use him in the way the New World expected. There had been speeches, whistle-stop tours, press conferences and parades, all of which Einstein had taken with a smile and a bow and a few words in German – and the result had been a triumph. Einstein had emerged as a spiritual leader of the Jews. When he had arrived with Weizmann at Cleveland, Jewish businessmen had closed their establishments so as to be able to join in a parade of solidarity to the City Hall.

The peaceful demonstrations and gestures of trust which Einstein had found on his foreign trips were, however, interludes from the realities of life at home: Germany. There the political scene was still festering, with violence never far from the surface. The Foreign Minister of the day was a Jew, Walther Rathenau, whom Einstein knew as a friend. There were qualities in the man of which Einstein was suspicious: he had a weak vanity and an obvious admiration of the spirit of Bismarck and the Junker military mentality; this love of Prussianism shown by some German Jewish intellectuals was a paradox which Einstein could never understand, just as he could never understand why some German Jewish refugees, after having lived many years in America, still yearned for their old Fatherland. There were, however, qualities in Rathenau which he considered worthwhile. He liked his eloquence and volubility and his earnest desire for international understanding; he particularly approved of the absence of narrow-minded nationalism from his political dealings; Einstein's controversial visit to Paris in March 1922 had been at the advice of Rathenau.

On 24th June 1922, Rathenau was assassinated by a group of reactionary students: it was a piece of brutality which suggested that preparations were being made for a right-wing revolution. Rathenau, a Jewish politician with suspected Bolshevik sympathies, had been an obvious target for violence. Einstein was now asked whether he thought a memorial service for Rathenau should be held at the university. He agreed that it should since he considered that silence on the part of the university could be interpreted as an expression of sympathy for Rathenau's political opponents.

Einstein's attitude was duly noted. A rumour began to spread round Berlin that Rathenau's murderers had him next on their list. It was not merely that Einstein was a Jew and a supporter of Rathenau which made him so vulnerable. Prejudice had spread to his physics, and right-wing politically minded scientists conveniently classed relativity theory as Bolshevik nonsense. Einstein wrote of his troubles in a letter to the gentle Prussian physicist Max Planck:

> . . . A number of people who deserve to be taken seriously have independently warned me not to stay in Berlin for the time being and, especially, to avoid all public appearances in Germany. I am said to be among those whom the nationalists have marked for assassination. Of course, I have no proof, but in the prevailing situation it seems quite plausible. . . .

The trouble is that the newspapers have mentioned my name too often, thus mobilising the rabble against me. I have no alternative but to be patient – and to leave the city. I do urge you to get as little upset over the incident as I myself.[2]

But Einstein made no withdrawal from public life. His name was now in the newspapers and he tolerated it there so long as it referred to one of the causes which were dear to him. For the rest of his life, and with few reservations, he was prepared to lend his name to any cause professing to further peace and internationalism. He was, of course, drawn into organisations on the highest level; in 1922 he was invited to join the League of Nations Committee on Intellectual Cooperation. Since this was four years before Germany's admission to the League itself, there were problems. Some of the French on the Committee objected to having a German joining, and some of the Germans objected that Einstein was not a German but a Swiss Jew. It was the old parable in different circumstances.

Even the award of the Nobel Prize to Einstein was not kept free of an unpleasant political taint. By the time it was made in 1922 Einstein was already acknowledged as one of the great physicists of his time, so that the Swedish Academy can scarcely be accused of having been hasty in his case. Relativity theory and Einstein's name, however, had been linked with so many controversies of a political nature that the Swedes were afraid of opening their Award Committee to a charge of political alignment. The citation was an unfortunate compromise; it referred back to the work on photo-electricity which he had done in 1905 and generalised the rest. The award was made to Einstein for 'the photo-electric law and his work in the field of theoretical physics'. Relativity got not so much as a mention.

The experiences of the decade following the Great War saw the beginnings of a considerable change in Einstein. Scientifically these were fallow years; they were the years of financial problems: the result of inflation in Germany during a period when he had to provide for a now estranged wife and two sons. He had to take on consultancy jobs and use his knowledge of patents to advise in industrial legal actions simply to keep his family alive. His ideal of a life given over to theoretical science had had to be put in abeyance. This was not all; this was the period when he first had to live alongside men and women in cruel social conditions. In his own country he saw poverty and hardship intensified by streams of refugees from Eastern Europe. And he was now beginning to experience for the first time in his relatively sheltered life the harsh practices of anti-Semitism.

As a scientist with strong emotional motivation, Einstein was above all concerned with the intellectual beauty of his mathematical theories. To his engineer son he expressed the view that if a theory was not beautiful, then it was a bad theory. It was his hope that the relationships between human beings would also function best under conditions of beauty and fairness. That man's behaviour to man as he saw it during those years did not submit itself to any such laws, nor even begin to approach his ideals, worried him deeply. The first seeds of disillusionment were beginning to sprout.

His scientific fame gave him new opportunities to travel, and see conditions in the rest of the world, and he took these willingly enough, but always keeping a wary eye on what was going on back in Germany. He was delighted when his ship pulled alongside the quay at Shanghai and German schoolchildren turned out to sing a song of welcome to him, though their song, *Deutschland, Deutschland über alles*, could have done little other than give a quizzical edge to his smile.

When he was gushingly welcomed by his countrymen in Argentina he was under no illusions as to the ambiguous nature of his esteem and he wrote, 'Strange people, these Germans. I am a foul-smelling flower to them, and yet they keep tucking me into their buttonholes.'[3]

There were even occasions when Einstein was an out and out embarrassment to his colleagues. In 1927, when anti-German passions were still running high in France, a centennial meeting in honour of the French physicist Fresnel was held in Paris. Each German who entered the meeting was politely but coolly received. When Einstein entered, the gathering of scientists broke out into a round of spontaneous applause. It was clear that the respect in which he was held was above considerations of his nationality.

Now, as he entered middle-age, his suspicion of Germany and her motives was as strong as it had been when he was the insignificant clerk in the Swiss government office. But still his suspicion had not yet turned to fear. It was the late 1920's and the platforms for Peace were innumerable. Einstein appeared wherever he felt he could make a valid contribution, as an author, as an orator, as a signature, as a scientist and where necessary as a German. But the suspicion he felt for Germany was no less than the suspicion Germany felt for him. Amongst his intellectual colleagues were some who in 1914 had been happy to link German culture with German militarism and for them to have to watch his trips abroad being treated with ambassadorial respect was a bitter pill to have to swallow. Increasingly, both he and his science were becoming subjects for attack by the German right-wing press. He took no trouble to reply. Whatever deep-felt protest he wanted to voice he kept for his condemnation of war and armament. He was unambiguous and uncompromising. Just as in 1920 he had said: 'My pacifism is an instinctive feeling, a feeling that possesses me because the murder of men is disgusting'[4] – so in 1929 he was saying:

'I would unconditionally refuse all war service, direct or indirect . . . regardless of how I might feel about the causes of any particular war',[5] and in 1930, 'War seems to me a mean, contemptible thing: I would rather be hacked in pieces than take part in such an abominable business.'[6]

This, of course, was the customary language of pacifism, but for Einstein these were meaningful words. He had committed himself and he had chosen what he considered to be the appropriate language. He spoke the words which many hoped would be spoken by the most famous living scientist. He would lend himself to war at no price. His reputation as a scientist had put him in the position where he was able to make a strong moral judgment and, though it might alienate him from the country where he lived and worked, he had not shirked his duty as he perceived it. The conscience of the scientist was clear, and clearly applied.

In the early years of the 1930's Einstein continued to travel and to speak for peace and disarmament. He did the Grand Tour: Geneva, Paris, London and American cities. The pattern repeated itself: reception, peace meeting, interview, polite applause and the stares of curiosity. It was in America that the possible futility of what he was doing struck him most forcibly. It was not just the newspaper reporters and their joke questions on relativity, or their requests for his views on prohibition or enquiries after his belief in God, which in the end he began to find depressing: it was the lack of reaction of his audience. He found, perhaps too often, that the introduction to a speech which he considered deadly serious would be made by a chairman who wanted only to crack a few light jokes, and that the speech itself would be politely tolerated by an audience struggling against boredom. They were, he believed, a propertied class for whom his words could have no meaning. Perhaps he was unfair to his audiences who often failed in the struggle to keep awake during the sheets of speech read by the quiet tousled-haired orator, speaking in a language they did not understand, followed by an often inadequate translation in a middle-European accent.

The crowds, nevertheless, turned out to watch, if not to listen and the magnetism of the man was still tremendous. Even if he himself had some doubts about the value in hard terms of realities of his own efforts for peace, George Bernard Shaw had no doubts about Einstein's position in the league table of men. Shaw, on one occasion in the Albert Hall, said, to cheers from the audience:

> . . . And supposing that I had arrived here tonight to propose the toast of Napoleon. Well undoubtedly I could say many very flattering things about Napoleon, but the one thing which I should not be able to say about him would be perhaps the most important thing and that was that it would perhaps have been better for the human race if he had never been born. Napoleon and other great men of his type, they were makers of Empires: but there is an order of men who get beyond that. They are not makers of Empires but they are makers of Universes and when they have made those Universes their hands are unstained by the blood of any human being on Earth. Ptolemy made a Universe which lasted for 1,400 years; Newton also made a Universe which has lasted for 300 years; Einstein has made a Universe and I can't tell you how long that will last.[7]

Shaw's references to Einstein's metaphysical creations were perhaps highflown but they undoubtedly indicated the great esteem which the world held for this scientist. The Institute of Intellectual Cooperation of the League of Nations judged him equally highly and in July 1932 gave him the opportunity of debating in an open letter, with any individual he cared to choose, any problem of his own selection. Einstein chose Sigmund Freud as his correspondent, and the problem of delivering mankind from the menace of war as the issue to be discussed. Of the half dozen or so minds which have most influenced 20th century thought, the League of Nations now had two of these working at its most pressing problems.

In the past between these two men there had been mutual respect, but a certain lack of mutual understanding. Einstein had reservations about Freud's work, but at least admired his literary style, whilst Freud said of Einstein, 'he knows no more about psychology than I do about mathematics'. Nevertheless, the correspondence went ahead. Einstein wrote asking Freud to apply his psychology to control man's 'psychosis of hate and destructiveness'. Einstein believed that his own science was impotent when it came to the control of peace, and therefore had to ask Freud whether his science could be applied to the solution of the problem. Yet, in a few years' time, it was to be Einstein's science, physics, and its unforeseeable applications, which was to be used to attempt to control war and effect a peace.

Freud replied and the correspondence was published, but not before the events in Germany had begun to threaten with violence the lives of the two Jewish letter writers, as they had the lives of millions of other Jews. It was a supreme irony that whilst Einstein's letter was being prepared for publication in Germany in the spring of 1933, his summer home near Berlin was being raided by Nazis on the pretext that arms were hidden there.

When the crowd of Nazi Storm Troopers, shouting and waving their flaming torches, stamped their way through the streets of Berlin towards the Reichstag, Einstein was in America doing scholarly work in California. When his summer home was ransacked by jack-booted young men he was on board ship on his way back to Europe. The wave of terror continued to spread through Germany. Göring had already taken command of the Prussian police and had ordered his men to shoot without fear of consequences; in the 'Brown Houses' the enemies of the Nazi party – Communists, Social Democrats and Jews – were being beaten up, and political murders were never more acceptable. Pacifists were officially considered to be enemies of the state.

The concentrated attempt to remove by any means whatever all Jewish influence on German culture was now intensified. Even physics, and particularly theoretical physics, was divided into Aryan and non-Aryan science, and what was non-Aryan was suspect. Einstein was attacked in speeches and in print; in the

Völkische Beobachter, Hitler's most influential paper, German scientists were reproached, not only for having accepted relativity theory, but for having failed to see how wrong it was to accept Einstein, 'this Jew, as a good German'. It was assumed that the purity of pure science could only be assured if work was done in a completely Aryan environment and that if German scientists should become lost at any time in a theoretical wilderness then they should struggle out of it without recourse to guiding lights from Jewish natural philosophers. The tone was set by the new Minister of Education, Bernhard Rust, who said, 'National Socialism is not an enemy of science, but only of theories'. But the attempt to eradicate so-called Jewish theories from German universities met with some difficulty. Intellectually honest physicists had either to accept relativity or flounder; therefore, when they taught it to students they did so without mentioning the name of Einstein. This attitude was to persist throughout the Hitler régime and some of Germany's most brilliant scientists were to make the awful compromise. They must have known of the dishonesty of their attitude and can only have maintained it for one of two reasons: that of self-preservation, or that of the preservation of science in Germany. One of Germany's most respected young theoretical physicists, Carl von Weizsäcker, the son of Hitler's Secretary of State, faced this particular dilemma with little apparent credit. A document in his handwriting and captured before the end of the war followed approved racial-scientific phraseology by referring to 'the Jewish followers of Einstein'.[8] Weizsäcker's unpleasant indecision is shown by the crossing out of this and other passages followed by some indications of afterthought as to whether they should be allowed to remain.

Einstein did not make the mistake of returning to Germany. On 28th March 1933 he reached Europe and went to Le Coq-sur-mer in Belgium. There he knew himself to be sure of at least a temporary refuge. For several years he had had an informal friendship with King Albert and Queen Elizabeth of the Belgians and on several occasions the Queen had played second violin to his first in quartets. He knew that whilst there he could count on Belgian protection, though how much Belgium could count on its own self-protection nobody at the time knew.

Einstein's immediate problem was the severing of relationships with the country of his birth. It was not easy for him. In Berlin he had shared the life of a group of scientists of high intellectual capacity. Many of these men had earned his respect and stimulated and encouraged his work. One of these, one of the most distinguished, whose contributions to quantum physics will be remembered far beyond this century, was Max Planck. It was Planck who as a slim Prussian aristocrat, along with Walter Nernst, had travelled to Zürich in 1913 to persuade Einstein to leave Switzerland for Berlin. And it was Planck who was now the foremost member of the institution which had taken in Einstein as an honoured member; the Royal Prussian Academy. If Einstein had not offered his resignation Planck would have been bound to ask for it. Einstein saved him the discomfiture and embarrassment. Planck, the man who believed that the Hitlerian oppression was a temporary state of affairs which would quickly pass, and who later was personally to plead the cause of Jewish scientists to Hitler, was to finish his days as a broken old man with his family, his country and its science shattered.

There was an acrimonious exchange of letters between Einstein and the Prussian Academy. The Academy welcomed his going with a published statement:

> We have no reason to regret Einstein's resignation. The Academy is aghast at his agitational activities abroad. Its members have always felt in themselves a profound loyalty to the Prussian state. Even though they have kept apart from all party politics, yet they have always emphasised their loyalty to the national idea.

The Academy further wrote reproving him for not having taken Germany's side against the barrage of vulgar criticism of National Socialism from abroad. Even Planck obviously believed the German press reports that Einstein had indulged in atrocity-mongering. The Academy suggested that a kind word at the right time from such a famous man as he might have done Germany a power of good. Einstein knew it only too well and his reply was sweepingly uncompromising in its condemnation of the treatment of the Jews in the new Germany.

> ... By giving such testimony in the present circumstances I should have been contributing, even if only indirectly, to the barbarisation of manners and the destruction of all existing cultural values. It was for this reason that I felt compelled to resign from the Academy, and your letter only shows me how right I was to do so.[9]

This was the second time that Einstein had made a break with Germany, but on this occasion there was no going back: he never again entered the country. It was, however, a turning point in his life in more ways than one. Although he found that he could be uncompromising in his attitude towards Germany, he could only maintain this attitude at the expense of his most coveted cause: the cause of peace. For years now he had used the position of fame which science had given him as a pulpit for his pacifist ideals. Now, as a refugee Jew, sitting in one of Europe's smallest countries bordering a rearming Germany in a mood of expansion, he knew that he had to compromise his ideals. It was the beginning of a new and fateful phase.

In June 1933, a French pacifist named Alfred Nahon made a request to Einstein which he had no reason to believe Einstein would deny. Two Belgian conscientious objectors named Day and Campon had refused to do military service, had been arrested, and had been imprisoned in Brussels. Nahon made a passionate plea to Einstein to speak in their defence at their trial. Even if he were not willing to appear in the court Nahon could reasonably expect a message of sympathy or a plea for their right to refuse to fight. After all, Einstein was the man who had said: 'I would unconditionally refuse to do war service, direct or indirect, and would try to persuade my friends to take the same stand, regardless of how the cause of the war should be judged.'

But conditions had apparently changed, and he was now judging the cause of the threatened war by different criteria from the ones he had used when he wrote these words. He was well aware of the nature of his volte-face when he wrote to Nahon: 'What I shall tell you will greatly surprise you. . . . Were I a Belgian, I should not, in the present circumstances, refuse military service. . . .'[10]

He was saying that he would not speak for Day and Campon and was recommending that Belgium should get ready to fight. Einstein's view was that the only way of defeating militarism under the existing conditions was to oppose it by force. Einstein himself, as well as many who have since tried to explain away his attitude, claimed that he was not surrendering the principle of pacifism for which he had stood for so long, but merely putting it in abeyance. It was playing with words. Even Hitler could have argued that he intended to fight only until he chose to stop.

What Einstein had done was to look around and to take in the warning signs before many others who were in a better position to see. He saw what he should do, not as a scientist, but as a public figure whose decision would be widely noted. The decision he made caused him much heart-searching but it was proper and wise. It needed no rationalisation. That Einstein had recommended that young Belgians should stand up and fight for their country was reported in the newspapers of the world.

Not unnaturally the reaction among pacifists was at first one of amazement and almost disbelief. Lord Ponsonby, a convinced pacifist, wrote to Einstein asking him not to publish any more of his views since 'every chauvinist, militarist and arms merchant would delight in ridiculing our pacifist position'. Then the reaction turned to bitterness. One official of the organisation known as the 'War Resisters' International' felt bound to point out that '. . . this failure does not mean that our principle is unsound but that humanity is, after all, very weak'. To his diary Romain Rolland confided: '. . . It is quite clear to me that Einstein, a genius in his scientific field, is weak, indecisive and inconsistent outside it. I have sensed this more than once. . . .' And the letters to Einstein poured in.

He had decided to settle in America. *En route* he visited England where he talked to several British politicians about the situation in Germany as he saw it. He was also obviously seeking reassurance that some sort of early action would be taken. He went away from a visit to Winston Churchill in a spirit of some optimism: 'He is an eminently wise man; it became very clear to me that these people have made their plans well ahead and are determined to act *soon*. . . .'[11] *Soon*, however, was still several years away.

The position he had accepted in America was at the newly established Institute for Advanced Study at Princeton. In many respects the life it offered was very similar to the one he had expected when, twenty years before, he had moved to the Physics Institute of the Kaiser Wilhelm Gesellschaft in Berlin. Here again he had the promise of a nucleus of stimulating minds, no duties other than the research he chose to do, and as much or as little contact with students as he felt he needed. There were also advantages which Berlin had not offered. The new Institute at Princeton was not only richly endowed with dollars, it was richly endowed with a liberal intent. Admission to the Institute for research workers was to be based entirely on ability with no account whatever of sex, race or religion. There was also another attractive factor which as a young man he had not needed, but which now, as a wandering middle-aged Jewish refugee, was comforting: the tree-lined campus of quasi-Gothic greystone buildings growing next to the small town of white-painted weatherboard houses gave to the place a cloistered atmosphere of security, and of shelter from the colder political winds of Europe.

The acceptance of this safe haven was Einstein's withdrawal: it was the acceptance of the inevitability of disillusionment in life. It was also the intensification of the change in him which had begun in the period following the Great War. Just as he was now beginning to recognise that a flawless and beautiful formula for the relationships between man and man was an unrealisable dream, so he began to suspect that the grand design of his latest scientific work would always stay out of his reach.

This work, with which Einstein was to occupy himself for the rest of his life, was an attempt to find a 'Unified Field Theory': a theory which would generalise gravitational phenomena that he had dealt with in his earlier work, and which would also take into account electromagnetic phenomena. It was a mammoth intellectual task. Einstein never considered himself to be a good mathematician and was despondent that the mathematical tools for the job might evade him. The work, begun in the 1920's, was still incomplete at his death.

Many believe that these later years, spent cloistered in Princeton, were wasted years, though Einstein realised the possibility of never being able himself to bring his self-appointed task to a conclusion.

The fact that Einstein had to acknowledge that his designs of beauty for mankind and for science were merely tantalising but unattainable aims ended the first phase of his life. The qualities of the aggressive and emotional revolutionary had retreated into the background with his withdrawal. The face which the disillusionment now left on view to the world at large was that of the benign and tolerant physicist.

Nevertheless, one persistent attitude which he maintained and still voiced strongly from his life of seclusion was his passionate mistrust of Germany, now turning to fear; still he was speaking of the Germans as having been 'drilled into slavish submission, military routine and brutality'. On several occasions he tried to warn Americans of the dangerous game of isolationism which some were playing, and the need for collective security against Hitler.

But the apparent futility of his anti-fascist pamphlets and his anti-war speeches was pressing in on him and emphasising his impotence. His increasing despondency was causing him to feel his age: he had been the elder statesman of science for many years now. Yet he was by no means an old man and was scarcely 54 when Hitler came to power. In February 1935, he wrote to Queen Elizabeth of the Belgians:

> ... Among my European friends I am now called 'The Great Stone Face', a title I well deserve for having been so completely silent. The gloomy and evil events in Europe have paralysed me to such an extent that words of a personal nature do not seem able to flow any more from my pen. Thus I have locked myself into quite hopeless scientific problems – the more so since, as an elderly man, I have remained estranged from the society here. . . .[12]

Solitude, he believed, developed the personality and built up in the individual a resistance towards mass suggestion. In this isolation, as he told the Belgian Queen, the horror beginning in Europe seemed part of some other existence: only the stories the newspapers were telling and the frightening letters from Europe left him with a feeling of guilt.

But Einstein was not allowed to keep his isolation sacrosanct. The problems of the world were brought to his doorstep by the short ebullient figure of the scientist who in the 'twenties had once been out of a job and had asked his advice: Leo Szilard.

A story is told that Leo Szilard was once asked to sit on a murder trial jury. When the time came for the jury to withdraw and consider its verdict a long and heated discussion took place leading to a split verdict: eleven men found the prisoner guilty; one, Szilard, unmoved by the forceful majority, found him not guilty. Hearing this the judge directed the jury to retire again and to reconsider its verdict. In the jury room an even more protracted and animated session followed with jurists arguing as a group and in a succession of small groups. In the end the twelve good men and true returned to the courtroom still with a split verdict: now eleven found the prisoner not guilty; one, Szilard, found him guilty.

The story might be apocryphal, but it serves to illustrate the influence of the diverse workings of the brain of the quixotic Hungarian.

Of the scientists who, in the first few months of 1939, were looking at the possibilities of fresh and vastly important consequences of newly discovered uranium fission, Leo Szilard was by no means the most significant. He was one of many physicists who carried out experiments during what was later to be realised as one of the most intense periods of creative scientific research of the century. But it was Szilard whose bubbling mind saw new and often strange paths along which the results of scientists' work might lead mankind. He was constantly alive to the responsibility of scientists for the consequences of their work, and his own expressions of this responsibility often took unexpected and unconventional forms.

F

Nuclear physics, which had attracted his and so many other scientists' attention, was suddenly bursting with an amalgam of ideas and a ferment of laboratory experiments. Since 1932 when Cockcroft and Walton at the Cavendish laboratory in Cambridge demonstrated that Einstein's relationship between mass and energy, $E=mc^2$, held precisely in an experiment, it had been realised that the time might not be far away when utilisable quantities of energy could be got from the atomic nucleus.

On 16th January 1939 the distinguished Danish physicist, Niels Bohr, arrived by ship in New York. Waiting to meet him at the quayside was the Italian émigré, Enrico Fermi. Both these men had already made contributions to the knowledge of the structure of the atomic nucleus, and would make still more. One of the subjects which they were most keen to discuss that day was the work which had been carried out a few months before in Germany in the laboratory of Otto Hahn. His experiments had involved the bombardment of uranium atoms by neutrons. Just before Bohr had sailed from Copenhagen, he had listened to a theory advanced by Lise Meitner, a Jewish refugee who was Hahn's co-worker, and her nephew, Otto Frisch, which would explain Hahn's results. The theory described in detail how the uranium atom could be split into two smaller atoms with the release of relatively large amounts of energy. A few days after Bohr's arrival in New York he received a telegram from Frisch telling him that an experiment had now been carried out which confirmed the theory.

Both Bohr and Fermi realised the immense significance of this news of what had now been dubbed uranium fission. If a neutron, in splitting a uranium atom, caused more neutrons to be emitted then there was the possibility of a chain reaction being started. Since the fission of each atom of uranium caused a small amount of matter to be converted into energy in accordance with Einstein's equation, then a few pounds of uranium could, in theory, have the explosive and destructive power of many thousands of pounds of dynamite. Before the end of that month both Bohr and Fermi were speaking at a Washington conference on theoretical physics and spreading the latest news to American scientists, as well as to the press agencies of the world.

And so, in spite of the obvious political uncertainties of the times, the menacing attitude of the Reich and the absolute sureness that German scientists were heavily involved in these latest developments in physics, the belief that a scientist's knowledge should be common to the international brotherhood of science was being as widely practised as it ever had been. After Bohr's announcement, the activity in the laboratories of the world stepped up, every nuclear physicist and chemist wishing to contribute – and publish – his part in this great burst of knowledge. It was from Frederic Joliot's laboratory in France that the announcement came that an experiment had been performed which showed that during the fission of a uranium atom by a neutron, more neutrons are released. This was a vital piece of evidence which physicists needed in order to establish that, in principle at least, an atomic bomb could be made, though at the time no one had thought of coining this convenient name for the end product.

After the Washington meeting Niels Bohr went to the Princeton Institute for Advanced Study to carry out some theoretical work on the new developments with his one-time student, John Wheeler. Fermi returned to Columbia University, where he was Professor of Physics, to carry out experiments on the uranium chain reaction; working with Fermi was Leo Szilard. Fermi and Szilard, like Einstein, were typical of the many intellectuals who had fled, who had been chased out, or who had simply walked out of Europe in the 1930's. Most of them, like Einstein and Szilard, were Jews, or like Fermi, whose wife was Jewish, had enough reason to be wary of the mounting racial intolerance in the countries in which they lived and worked. Fermi was an Italian who had used Sweden and the award of his Nobel Prize as his family's convenient escape route to America. At the ceremony Fermi had refused to obey instructions from Italy and had not given the fascist salute to the King of Sweden: instead he had shaken hands. His little gesture of disapproval of Mussolini's fascism had not been well received in Italy, where it had been condemned as 'un-Roman and un-manly'. He had no reason to fear this or other petty expressions of displeasure: he had an academic job waiting for him in the United States. Szilard was Hungarian, but had travelled widely enough in other Western European countries to realise the dangers of the winds of fascism. In New York he had no official post but was working with Fermi as a guest of Columbia University.

As a group it was the refugee physicists who most clearly not only saw, but feared, the consequences of their own work. When the speculation of the future of their researches was at its height in the middle of 1939, Fermi discussed his fears with a fellow physicist, the Dutch refugee, Sam Goudsmit. To emphasise his point more dramatically Fermi took Goudsmit to the window of the tall New York building where they were talking and with his finger pointed out a huge district of the city. This was the area he believed could be destroyed by one atomic bomb if his predictions of its potential were correct.

Already by March of 1939 Fermi had talked to representatives of the United States Naval Research Laboratory and the Army's Bureau of Ordnance to alert them to the possibilities of a new weapon, but he had to emphasise the uncertainties. Fermi was as aware as any physicist of the practical difficulties which lay between the scientific principle and a working bomb. Szilard too had had informal discussions with a technical adviser to the Navy when they met at an American Physical Society meeting in Princeton in June; but if the Navy was taking the subject seriously, then they did not confide in Szilard, nor did they offer to finance his work with a government contract.

Early in 1939, at the same time as uranium fission was being most intently investigated, Hitler was marching into Czechoslovakia, and Czechoslovakian mines were a well-known source of uranium ore. There was no doubt whatever that German scientists knew what the results of a chain reaction in uranium might be – it was, after all, from German laboratories that much of the impetus for this nuclear research had come. Szilard was well aware of all these facts and his experience of Europe told him that if Hitler's physicists could turn the principle of atomic energy into an horrendous weapon then there was no knowing where the tentacle of fascism might spread.

It is not really surprising that Szilard's concern for mankind should have been listened to with only politeness and without the promise of action from an informal representative of the United States' Navy. Szilard, after all, had only the most casual of connections with the United States. Firstly, he was an alien, and secondly, at Columbia where he was working, he was nothing more than a guest in Fermi's laboratory without any university appointment. He was nevertheless an enthusiastic guest and had even gone so far as to borrow money in order to get hold of some radium to

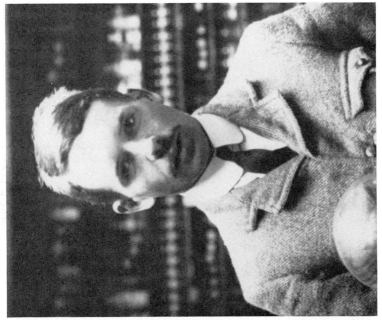

2 H. G. J. Moseley in 1910

1 Alfred Nobel

3 E. T. S. Walton, Lord Rutherford and J. D. Cockroft in 1932

4 Otto Hahn and Lise Meitner in 1959, at the opening of the Hahn-Meitner Institute for Nuclear Research in Berlin

use as a neutron source for the laboratory's experiments. His enthusiasm for action was as great outside the laboratory as it was inside. By now he was convinced of the real possibility of a race to make an atomic weapon in which the scientists of Fascist Germany had short odds. In addition to the important source of uranium in Czechoslovakia, Germany if necessary could get more large quantities of the ore from the Belgian Congo. Szilard was intensely concerned that Belgium, a country nervously contemplating the possibilities of invasion by Germany, might unwittingly be in the process of selling a potential high explosive to her future enemy. He talked over the dangers with another émigré physicist, Eugene Wigner, and of how they could get the information to the Belgian Government.

It was at this point that it occurred to Szilard that a smooth and unofficial diplomatic contact with Belgium might be made via Einstein. He knew of Einstein's friendship with Queen Elizabeth of Belgium and that they frequently exchanged letters of greeting. Here might be a reliable and respected contact to warn the small nation. But for this chance recollection of Szilard the name of Einstein might never have been associated with the atomic bomb, other than as the originator of the equation relating mass and energy. Now Szilard was implicating him in developments which, for the duration of the war at least, he would otherwise have been completely unaware.

Szilard and Einstein each had considerable respect for the other's opinions. One friend recalls that one of Szilard's all-embracing characteristics was to treat all men as equals with the exceptions of Albert Einstein and James Franck. And a few years earlier Einstein had written: 'I consider Szilard a fine and intelligent man who is ordinarily not given to illusions. Like many people of that type, he may be inclined to exaggerate the significance of reason in human affairs.' But the time had come when Szilard, convinced of the significance of his reasoning, now sought action.

It was mid-July and Einstein, as he frequently did, was spending the summer away from Princeton at Nassau Point on Long Island where he could sail and spend some time in the open air. Szilard later recalled that when he and Wigner at last found Einstein and the news was broken to him that there was a possibility of a chain reaction in uranium, Einstein replied, '*Daran habe ich gar nicht gedacht!*' – 'That had never occurred to me!' If Szilard's recollection is correct, then this reply is a surprising comment on the extent to which Einstein had allowed himself to become cloistered from the latest developments in physics. It was now six months since Niels Bohr had arrived in New York with his news of Hahn and Meitner's fresh work, and from New York he had travelled within a few days to the Institute for Advanced Study at Princeton, where Einstein worked. Not only physicists talked freely amongst themselves about what might be the consequences of a chain reaction: some newspaper correspondents were already beginning to speculate and ask questions. Yet Einstein, according to Szilard, was quite unaware of the scientific furore.

When Szilard and Wigner told their fears about the Belgian Congo sources of uranium ore Einstein immediately agreed that some steps should be taken to keep it out of the hands of Germany. Of all scientists, he was one to whom suspicions of that country's motives need not have been emphasised. Only the method of warning Belgium needed some discussion. Einstein felt it better not to dictate a letter to his friend the Queen, but to somebody appropriately below her level in the Belgian government. This, however, presented the problems of communication with a foreign government and Wigner suggested that the U.S. Department of State should be consulted. He also suggested that at the same time it might be wise to suggest that the U.S. look to her future and stockpile uranium. Before the day was out all three men had turned over in their minds and discussed the rights and wrongs of advising the American government to take part in the research which might lead to an atomic bomb.

Whether or not this group, meeting in the middle of July of 1939, altered the course of history, whether they brought forward the manufacture of the atomic bomb by months, weeks, days, or not at all, cannot be assessed. There is little doubt that, had they not decided to take action of some kind, the task would have been taken on sooner or later by others in America just as keenly aware of the importance of the new discoveries as they were. And it is highly probable that the composition of any other group which might have chosen to discuss these matters at that time must have had similar elements in its composition as this hybrid trio. This might at first seem unlikely; the collective pedigree of the group was extraordinary, if not extreme. Einstein, Szilard and Wigner were none of them native born: none had even spent their youth in the United States; Szilard was still an alien and Einstein still held his dual German-Swiss citizenship; Szilard and Wigner were natives of Hungary and Einstein of Germany and, indeed, the language which they used to plot how best to defeat the ends of Germany was German; all three were Jews and all three were too well aware of the meaning in real terms of Nazi fascism; all three were much respected and widely counselled physicists of unusual abilities. In short they were men with intimate knowledge of Europe, with specific reasons for fear, and with an insight into the still generally underestimated possible applications of science to war. As history later showed, other men of these same experiences and qualifications were soon considering similar action in other parts of the world.

Szilard and Wigner left Long Island without any firm course of action having been taken. Szilard by this time, however, had decided to investigate another channel of communication to higher authority. At the suggestion of a friend he visited an economist named Alexander Sachs who was reputed to have established some informal means of communication with President Roosevelt. It was Sachs who was the first non-scientist to recognise the true significance of Szilard's warnings and suggested that rather than the Belgian Royal Palace, the White House ought to be the destination of the scientists' letter. He was also convinced of the importance of the eminence of the name to be put at the foot of the letter if it were to influence Roosevelt: Einstein's.

Once again Szilard set out for Einstein's summer cottage, taking with him this time not Wigner, who was out of town, but another Hungarian: a young physicist called Edward Teller. Teller acted as scribe. It was not to be the only time that he was to be close at hand when events concerning nuclear weapons and their effect on mankind were discussed. Though there was no pomposity or sense of occasion about the affair. Einstein, as had now become his daily habit, was casually dressed and wearing carpet slippers as he dictated a draft letter in German. Within a few days English versions of the letter were prepared; the one Einstein signed read:

August 2nd, 1939

F. D. Roosevelt
President of the United States
White House
Washington, D.C.

Sir:

Some recent work by E. Fermi and L. Szilard, which has been communicated to me in manuscript, leads me to expect that the element uranium may be turned into a new and important source of energy in the immediate future. Certain aspects of the situation which has arisen seem to call for watchfulness and, if necessary, quick action on the part of the Administration. I believe therefore that it is my duty to bring to your attention the following facts and recommendations:

In the course of the last four months it has been made probable – through the work of Joliot in France as well as Fermi and Szilard in America – that it may become possible to set up a nuclear chain reaction in a large mass of uranium, by which vast amounts of power and large quantities of new radium-like elements would be generated. Now it appears almost certain that this could be achieved in the immediate future.

This new phenomenon would also lead to the construction of bombs, and it is conceivable – though much less certain – that extremely powerful bombs of a new type may thus be constructed. A single bomb of this type, carried by boat or exploded in a port, might very well destroy the whole port together with some of the surrounding territory. However, such bombs might very well prove to be too heavy for transportation by air.

The United States has only very poor ores of uranium in moderate quantities. There is some good ore in Canada and the former Czechoslovakia, while the most important source of uranium is Belgian Congo.

In view of this situation you may think it desirable to have some permanent contact maintained between the Administration and the group of physicists working on chain reactions in America. One possible way of achieving this might be for you to entrust with this task a person who has your confidence and who could perhaps serve in an inofficial capacity. His task might comprise the following:

(a) to approach Government Departments, keep them informed of the further development, and put forward recommendations for Government action, giving particular attention to the problem of securing a supply of uranium ore for the United States;

(b) to speed up the experimental work, which is at present being carried on within the limits of the budgets of University laboratories, by providing funds, if such funds be required, through his contacts with private persons who are willing to make contributions for this cause, and perhaps also by obtaining the co-operation of industrial laboratories which have the necessary equipment.

I understand that Germany has actually stopped the sale of uranium from the Czechoslovakian mines which she has taken over. That she should have taken such early action might perhaps be understood on the ground that the son of the German Under-Secretary of State, von Weizsäcker, is attached to the Kaiser Wilhelm-Institut in Berlin where some of the American work on uranium is now being repeated.

Yours very truly,
A. Einstein.

The signature, in a careful and simple handwriting, was that of the most widely known scientist of the century, and it was fixed to one of the most significant documents ever written by scientists. In that document, the first of its kind in the century, scientists were speaking not just as public figures whose success in science had given them sufficient eminence to be listened to: they were scientists who were speaking as scientists and by warning the President were acknowledging that they held responsibility for the consequences of their work. They were suggesting, as they saw it at the time, how their work should be best applied for the good of the community – if necessary as a weapon. The pity of it was that, of the scientists involved, only Einstein's name appeared on the letter. He knew only too well that its main task was to give the composition stature.

Long after the Hiroshima and Nagasaki bombs had been dropped, he was later to try to explain to a Japanese that he had been fully aware of the dreadful danger which would threaten mankind if the experiments which his letter to Roosevelt described should prove successful; the factor which had compelled him to sign was his lifelong suspicion of Germany, now turned to fear: if German militarists had in their hands an atomic weapon there was no telling what the consequences might be.

The result is that the document is known as 'Einstein's letter to Roosevelt'. The signatures of the other three scientists so far associated with the letter would have been just as appropriate as that of Einstein in terms of scientific familiarity, and probably more so in terms of authorship. His equation $E = mc^2$ cannot be denied as the starting point for a bomb which will turn small amounts of matter into huge destructive quantities of energy; beyond that, as Szilard discovered when Einstein confessed that

73

the possibility of a chain reaction had not occurred to him, he was out of touch with what his equation meant in terms of the experimental nuclear physics of 1939.

Fear of Germany had driven Einstein a long way from his position a few years before as an uncompromising pacifist. He had now gone so far as to commend a weapon of war to the President of his new country. It was a difficult decision, and if it was a happy one it was not to remain so for long.

Several weeks passed before Sachs judged the time to be right to bring the contents of the letter to Roosevelt's attention. By this time war in Europe had broken out, the general fear of Germany was heightened, and Szilard was sufficiently agitated at the delay to begin thinking about other possible means of reaching the President. There was no need. Roosevelt saw Sachs, heard him out and summoned his secretary, General Edwin ('Pa') Watson, with a short sentence that was to enter history: 'Pa! This requires action.'

The action which was taken was the appointment of an 'Advisory Committee on Uranium' which met on 21st October 1939. In addition to government representatives it was attended by Sachs, and the three Hungarians Szilard, Wigner and Teller. By 1st November the Committee was recommending the purchase of fifty tons of uranium oxide in its first report. At last action was being taken: at least, so it seemed at first. This early flurry by the Committee, however, did not develop into any movements which satisfied Szilard's burning wish for positive results. By February 1940 he and Sachs were once again calculating how best they could prod the Roosevelt Administration into doing something definite about uranium research. Their solution was yet again to persude Einstein to sign a letter intended for the President, though on this occasion not addressed to him. They invented the somewhat precious method of having Einstein address a letter to Sachs suggesting that he relay more information and warnings to the President. Einstein, as he most certainly realised at the time, was being used as a postbox for the White House. The letter was in two parts: the first again warned about the likelihoood of the dangerous progress of uranium work in German laboratories and the second was a naïve but successful piece of blackmail. It drew attention to the fact that Szilard was considering publishing his latest experimental results on a uranium chain reaction for all the world, including Germany, to see, and would do so unless Roosevelt considered the subject sufficiently important to suggest withholding publication.

This cheeky and dangerous attempt at manipulation of the President of the United States, which bears the stamp of Leo Szilard all over it, served its purpose well. Within a few days Roosevelt proposed an enlarged Advisory Committee on Uranium with Einstein, and any others he might like to suggest, as part of it. Not unnaturally, Roosevelt was unaware of the charade which was being played out for his benefit and had miscalculated Einstein's real role and intentions. Einstein sent his regrets for being unable to attend the meeting.

7: In Germany

If it were not for the two feed roads running through it and pouring heavy traffic on to the A.1, Godmanchester could be one of the best endowed of the little village-towns of England. As it is, the place is linked to Huntingdon by an umbilical cord of a never-ending stream of lorries. But during the war, when both petrol and traffic were in short supply, it managed for a few all too short years to regain its integrity. Then the half-timbered white cottages and the façades of the red-brick, Georgian houses returned to an isolation and quiet much nearer to the atmosphere which the original builders of the place had known.

Godmanchester, in July of 1945, was as unlikely a place as any novelist could conjure up from his imagination as the setting in which a group of German physicists should be gathered and should discuss the end of an era of German science. Certainly the townspeople were unaware during that summer of the distinction of the German visitors interned by American and British military authorities in Farm Hall in West Street of the town. The physicists included some of the most notable of the products of German Universities of the first few decades of the 20th century: Otto Hahn (the chemist much admired by Einstein), Werner Heisenberg, Carl-Friedrich von Weizsäcker, Max von Laue, and Walter Gerlach. They were men who had had profound influence on Germany physical thought; their contributions had helped to keep Germany among the leading scientific nations. They had each been captured in Germany during the preceding few weeks and brought to Britain in a group.

There were neither dangers nor discomforts for the interned

5 Einstein at Nassau Point, 1939

6 Victor Weisskopf, Marcus Oliphant and Leo Szilard at Pugwash, 1957

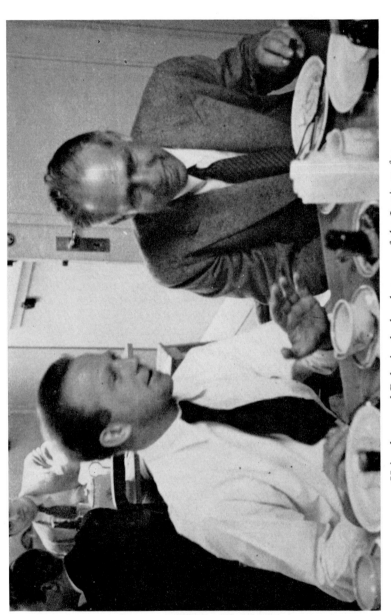

7 Heisenberg and Bohr in the lunch-room of the Institute for Theoretical Physics, Copenhagen, in 1936

Germans at Farm Hall. They had just come from a war-ravaged country and in most cases had spent their last working days under fire from enemy guns or bombs, they had been undernourished, and though they were still not sure of their future, it was no less unsure than the life they had left behind in their own country. Otto Hahn, for example, after a few weeks of Huntingdonshire food put on several pounds in weight; he was given a room on the first floor of the house with a view across West Street and through an ornamental avenue of trees to Port Holme, the meadow which Cobbett called, 'the most beautiful, and by far the most beautiful that I ever saw in my life'. It was the meadow of which Pepys, on a visit to his father in 1662, had written, 'the country-maids milking their cows there . . . and to see with what mirth they come all home together in pomp with their milk, and sometimes they have musique go before them'. In 1945 there were no local delights for the pleasure of music-lovers such as Heisenberg and Weizsäcker, but in the company of a guard the local cinema was not out of bounds.

Like all captives, no matter what the conditions of their imprisonment, the German physicists would have preferred their freedom and one of the first things they discussed as a group on arrival was whether they could build up a legal case for their release. They discussed which British scientists they might contact who would be sympathetic to their cause, and what the alternative to imprisonment in Britain might be. There was little else to do at Farm Hall other than to discuss their plight, to discuss each other, and to discuss each other's work. And it was the sure knowledge that if they were kept together then they would talk in this way, that was the reason for the disarming charm and simplicity with which they were confined. These highly intelligent men naïvely underestimated the motives of their captors. One of them, Diebner, soon after his arrival began to wonder whether there might not be microphones hidden about the tall Georgian rooms. 'Microphones installed?' replied Heisenberg laughing. 'Oh, no, they're not as cute as all that. I don't think they know the real Gestapo methods; they're a bit old-fashioned in that respect.'[1] Inevitably both question and answer were being carefully monitored and transcribed.

All of the physicists who had been interned had at some time or other during the past six years made some contributions to atomic physics and most of them had worked on projects which might have led to a German atomic bomb. The question was: how much did they know and what work had they done which Allied intelligence did not yet know about? The recorded conversations told the listening British officers all they wanted to know; but they were revealing in a number of other ways.

Not unnaturally the conditions of close confinement of the group bred a certain amount of short-temper; doubts and mutual uncertainties were soon expressed and in the most off-guard moments the physicists were critical of each other, of personal abilities, and of past motives. More, the conversations showed that not only were these men aware of the end of the political hey-day of the country of their origin, they were also at last aware of the end of an era of German scientific supremacy.

The outward proof of the quality of German science and technology during the first quarter of the 20th century is still obvious for all to see. A thriving dye-stuffs industry had been founded on the work of its excellent organic chemists; a powerful armaments industry flourished to such an extent that its tycoons survived two military and economic collapses which the armaments themselves had failed to prevent; and the esteem in which the rest of the world held its research and teaching universities was such that Germany was the natural Mecca through which all scientific pilgrims should unquestioningly wish to pass. Whereas today the European student invariably thinks of North America as the place to spend postgraduate or postdoctoral years, in the early years of the 20th century he would have gone (and with more difficulties and sacrifices than is the case today) to Berlin, or Göttingen, or Münich. This acknowledgement of German achievement still manifests itself in British universities and schools where students persist in presenting German as their compulsory 'scientific' language.

During the Great War there had seemed little doubt among representatives of German culture as to which standpoint the intellectual should take when an appeal is made to nationalistic sentiments, as was shown by the overwhelming support given to the *Manifesto of the Ninety-Three German Intellectuals*. But although such an eminent physicist as Einstein had refused to sign the document which insisted that 'German culture and German militarism are identical' there were no recriminations and few post-mortems of consequence among physicists themselves. German science continued to move towards a sure future of success under the guiding, shining lights of men such as the Prussians, Max Planck and Max von Laue, and under the Jews, Albert Einstein and Max Born. Among their pupils new stars were on the horizon and by the end of the 1920's there were enough young scientists and technologists to make the future appear to have no uncertainties in it.

On the surface the future of the armaments industry did not seem too rosy. In January 1920 the Treaty of Versailles had come into effect. It was a bitter pill to swallow, and not only for politicians. Although disarmament provisions had been made by the Treaty, army leaders did everything they could to circumvent these. By 1924 a highly secret 'Armament Office' had been established in Berlin to find out what munitions and equipment would be necessary to supply an army vastly in excess of that of the limits of the provisions of the Treaty. But besides the clandestine violations of the Treaty there were loopholes which could allow rearmament of a kind, and the German Army Ordnance Department began to look into these. In Germany as well as other countries both technologists and cranks were becoming excited by developments in a new field of science: rocket propulsion. The cranks far outnumbered the technologists. In those days science fiction had not reached the age of respectability. Any talk of launching a vehicle into space belonged firmly in this area of doubtful fiction, and the enthusiastic but under-financed amateurs who actually went so far as to dabble in the hardware of rocketry were little more than tolerated by amused, respectable scientists. Rocket propulsion was therefore as unlikely a field as any for a military organisation to decide to indulge itself in. But the loopholes in the armament provisions of the Treaty of Versailles were so small, and the alternatives to open weapon development so few, that the German Army Ordnance Department decided to finance a limited investigation into the possibilities. It was a throwaway bet by the Army of a little small change on a long outsider.

Army Ordnance began to look for a soldier with a technological background who could be attached to their ballistics section to look into developments of rockets as weapons of war.

The man they appointed was a young captain called Walter Dornberger. It was a fortunate choice. Dornberger was to stay in his job for more than 15 years and only to lose it because Germany, rather than his project, collapsed. In 1929 Dornberger was a young enthusiast not long out of the mechanical engineering school of the Berlin Technological University. By training he was a technologist, but by profession he was a soldier and throughout his career he never wavered from the dictates of his profession. The brief for his attachments to the Ballistic Council of Army Ordnance was clear and unambiguous: to produce a weapon of war, and this he set about doing with the enthusiasm of the young scientist and the single-mindedness of the young soldier. Dornberger was self-confessedly disinterested in politics. Like the rest of the dabblers in rocket propulsion, the amateurs, the cranks and the visionaries, he was excited by the prospect of space travel and of interplanetary communication, but he was never to waver from his task of building a rocket which could be made to carry a warhead into enemy territory. He was the unquestioning soldier administrator.

Dornberger's first move was to find collaborators: scientists and engineers with a more specialised technological background than his own, whose vision might well be as high-flown as the cranks, but of whose ability there could be no doubt. He began to employ reputable scientists with sound and conventional backgrounds, but he did not make the mistake of avoiding the amateur groups whose pathetic little projects were dying from lack of funds. One of these was the Amateur German Rocket Society, a group which had been carrying out experiments on a Berlin airfield. Dornberger found working there a young student who impressed him in a number of ways. The 19-year-old boy was tall, blond and good looking. He came from an aristocratic land-owning family which was not at all enthusiastic at his choice of a scientific career, but this attitude had not deterred him in the least. Most striking of all was the highly developed theoretical knowledge of this boy student, Wernher von Braun. Dornberger discovered that von Braun at the age of 17 had written out in long-hand on notepaper a 'Theory of Long Distance Rockets' based on Keppler's laws. His enthusiasm and his unquestionable ability were enough for Dornberger to disregard his lack of experience and by the summer of 1932 von Braun had been invited by the army to its experimental station to begin work on rocket combustion phenomena for a doctoral thesis.

That the army's and his own ultimate aims might not coincide was of no great concern to von Braun at this embryonic stage of his career: the means to the end were mutually exciting. Twenty years later he said:

> In 1932, the idea of war seemed to us an absurdity. The Nazis weren't yet in power. We felt no moral scruples about the possible future abuse of our brain child. We were interested solely in exploring outer space. It was simply a question with us of how the golden cow would be milked most successfully.[2]

Von Braun used the produce of the well-nurtured cow with skill and it was not long before his stature was growing in the eyes of the few other workers in the field. By 1934 he was lecturing at Berlin University on problems of liquid-fuelled rockets. By 1935 he was soberly warning the Ordnance Department not to expect too much after some successful tests in which rockets had reached altitudes of 8,000 feet. Already the Ordnance Department was thinking of a device with practicable military applications and a range of 30 miles. But von Braun was cautious. He wrote in a secret memorandum, 'The devices . . . represent little more than an interim solution without any refinements.' Nevertheless the young man of 23 had already committed his science, as had numerous other young men in the other leading technological nations, to the production of a weapon of war. The end was worthy enough: to produce a rocket which would be capable of free flight in space. The means were provided by the soldier's administrative ability to provide army funds. The immature von Braun was swept along by the opportunities which could only have been flattering to a young scientist.

Von Braun's vision was huge. Guided by Dornberger's senior engineering collaborators he was able, within the space of a few years, to foresee the scale to which future development would have to grow. Already the group had had modest successes with early models of rockets, but if reality and vision were to meet then available funds would have to rise to the grandness of the vision. One of Dornberger's chief contributions to the project was to maintain a continual hunt for money from any source; the more influential the visitors to the rocket experimental station at

Kummersdorf then the more grandiose the public relations show which Dornberger put on for their benefit. In March 1936 he invited General von Fritsch to view the three 3,500 lb thrust rating motors which had been developed. The invitation had been given with the sole purpose of soliciting for funds, and for the inviter and the invited the result was cheering. Von Fritsch was impressed by what he saw at Kummersdorf and Dornberger got his promise of funds. There was only one proviso, and it was one which he accepted readily: support would be guaranteed if the end product would be turned into a serviceable war weapon.

Results had been quick: so quick that the experimental rockets began to outdistance the range of the proving grounds, and by April 1936 Dornberger and von Braun were pushing blueprints of a new development station in front of General Kesselring, the Director of Aircraft Construction. The site for the new station had been found by von Braun during a holiday with relatives on the Baltic coast. For several months the rocket group had been looking for a spot which would be suitable for their experiments and at the same time which would not excite the interest of the local population or of casual visitors to the area. The place von Braun had found on the bay of Peenemünde had both the remoteness and the convenience. It was surrounded by pine woods on one side and bordered the small Baltic lakes on the other. The position on the sea would give the new station a flight corridor to the East North East of 250 miles along which rockets could be launched from easily camouflaged installations.

Kesselring's approval of the rocket group's work meant a more reliable supply of funds and Dornberger was able to buy the Peenemünde site for 750,000 Reichsmarks. This was a small sum compared with what was eventually to be sunk in the scheme; between 1937 and 1940 the project swallowed 550,000,000 Reichsmarks. By March of 1939 work on the A3 rocket ('A' standing for 'complete rocket assembly' or 'aggregate'), was so advanced that Peenemünde was considered ready for a visit from Hitler himself. For Dornberger, seeking top priority for the work, this was a crucial encounter and every effort was made to convince the Führer of the potential of the liquid-propellant rocket and to guarantee it a well-financed place in his armoury of the future. Dornberger outlined for Hitler the work which had been done since the transfer to Peenemünde and handed over the technical discussions to von Braun. Hitler's intolerance of inadequate factual explanations was well known, but on this as on future occasions, von Braun was an adequate advocate of his own schemes. In spite of this Dornberger was unsure of the impression created on his leader. Hitler had seen the static firing of a rocket, had seen the power, the colour and the flames, and had listened to the din of the roaring engines. All of this had been shown calculatedly to move his stamp of approval towards the vast nature of the scheme. Hugeness of purpose and of design had always peculiarly titillated Hitler's fancy; but although he was clearly impressed by the hardware of the rocket installation – the launch pads, the rocket motors and the metering arrays – he left Dornberger and von Braun unconvinced that he would be prepared to commit himself to the more expensive concepts which they had in mind and which would be necessary to develop their rocket into a production model. Goering's reaction to Peenemünde a few weeks later was very different. The public relations job which had failed with Hitler worked like a charm in his case. He was excited by all he saw.

Had Hitler's reaction to the rocket assembly been that of Goering's and had he given the project the top priority which Dornberger was seeking, there can be little doubt that the consequences to Hitler's enemies, who in five years' time would be seeking to invade Germany, would have been disastrous. Efficiently developed and protected rocket installations designed to bombard the Southern Coast of England could have ruled out any possibility of this area being used as a starting point for the Western invasion of Europe. But the visit of Hitler to Peenemünde was still six months removed from the time at which Germany and Britain were to be at war. The Peenemünde project engineers had worked their way through both failures and successes, and there were still innumerable technical obstacles to be overcome and which could be worked on without a top priority. There had been no precedent for their models and every step they had taken had been of a completely empirical nature. Even in their present state of knowledge of rocket development it was clear that theirs was a considerable scientific achievement. If, among the rocket scientists, there were any misgivings at the ultimate purpose to which their work would be applied by the military should it be successful, then there was no overt sign of it at this stage. War was on the horizon and the rocket scientists were prepared to throw in their lot to what they believed to be the common good. For several years von Braun had been experimenting with rockets as aircraft propulsion systems. On 6th July 1939 he signed a memorandum to the Reich Air Ministry for a 'Proposal for a workable fighter with rocket drive'.[3] It was a visionary paper years in advance of the technology of 1939. It saw the limitations of conventional rolling-start aeroplanes and suggested how propulsion power could be used to give a vertical take-off. Fortunately for Germany's enemies its proposals were too radical to be of use in the war which was just on the doorstep.

By the time war did come Dornberger, von Braun and the Peenemünde engineers still did not have the blessing and priority they needed. It was not from want of trying on their parts. They had advocated their science to the highest authority in the land – Hitler; and they had tried to carry him along with their own enthusiasm. They can never be reproached for a lack of confidence in either their own abilities or in the ultimate success of their grand design.

Even with limited resources progress continued. Within a few weeks of Britain's declaration of war two rockets had been successfully launched to a height of five miles and had been safely recovered after a parachute descent into the Baltic.

The young enthusiasts working in Germany on the problems of rocket propulsion during the ten years before the outbreak of war were engineers without a tradition to follow; their inexperience and immaturity were no barrier to what they were trying to achieve. But Germany's natural scientists of the 'thirties – her physicists and chemists – were walking in the footsteps of the giants of the trade. The strides which men like Albert Einstein and Max Planck had taken had caused physics to evolve in a new and exciting way which had not been foreseen at the turn of the century. Small men might have been frightened by such a tradition. But Germany was well endowed with young physicists and chemists who were able to pick up threads where the older men had left off. One of these was Werner Heisenberg. He had been a child when Einstein and Planck were publishing their greatest work at the beginning of the century, but by the time he was in his early twenties he was making contributions of the highest significance to physical theory. The Nobel Committee was in no doubt about his ability and by the age of 32 he had been awarded the Nobel Prize for his formulation of the famous Uncertainty Principle.

Heisenberg's contact with physics and physicists was by no means limited to the work going on in his own country. He had been stimulated by the visits of foreign scientists to German seminars and had even for a time worked in Copenhagen where he had come under the influence of the great Danish physicist Niels Bohr. He became one of Bohr's pupils, favourites and friends, though the friendship was to be peculiarly strained at a later date. But in the early 1930's, when the political relations of European nations were as teetering as they had been at any other time during the century, the internationalism of science was an established and readily accepted state of affairs. Whereas applied scientists had already been co-opted by German Army Ordnance to work on its secret weapons project, pure scientists had had to make no compromising decisions since their work seemed remote from any warlike application, certainly within the next few years. Physicists met, published, and discussed their work in one another's countries at will. But there were a few difficulties to be overcome for those who had nationalistic sentiments to battle with, and Heisenberg had his difficulties. His great love of his country was rooted in a pre-Hitlerian old-style nationalism, and he was not disconcerted by the beginnings of the new régime in 1933; indeed, he continued to work under it until its end twelve years later. Nevertheless the illogicalities of Hitler's racial policies, and particularly where these overlapped science, were troublesome to most non-Jewish scientists and unacceptable to a few of them.

Even before 1933, Jewish scientists had begun to trickle away from Europe to less insecure political climates. Einstein had left Germany in a flurry of publicity and others were to follow him with less attention paid to their movements. After 1933 the trickle swelled. These men and women, driven away by fear of Hitler, might have become some of Germany's greatest assets. Now the ardent Nazis among the scientists were the only spokesmen from the men of their culture whose voices could be distinctly heard. The message which such distinguished physicists as Philipp Lenard and Johannes Stark had to tell in their increasingly vicious speeches and publications was that what was Jewish could not be good physics. Einstein's theory of relativity was a case in point and became an obvious target for the unthinking fanatics. But Werner Heisenberg was not one of these. He clearly saw the stupidity of attempting to eliminate the theory of relativity from modern physics and in 1937 he published an article in the Nazi journal *Das Schwarze Korps* in which he set out to defend the validity of the theory. Nazi authorities did not take kindly to this implied criticism and a reply from Stark in the same paper labelled Heisenberg as a 'White Jew'. The attack worried Heisenberg, who had no wish to be out of step with the régime, and might even influence it for the good. Like other German scientists he had already trained with the peacetime army and had been a non-commissioned officer with the Alpenjäger, though the training involved had amused rather than excited him. To one friend he compared his military service to mountaineering complicated by the presence of sergeants.

Heisenberg's reaction to the personal slurs of *Das Schwarze Korps* was to attempt to take his defence of Einstein to a higher authority in the Nazi hierarchy and he wrote to a family friend, Heinrich Himmler, Chief of the Gestapo. The consequence of this correspondence was that Himmler had an investigation made of Heisenberg's personal loyalties. The report was eminently satisfactory and Himmler was to write on 21st July 1938:

> . . . I believe that Heisenberg is a decent man, who is still young and can still produce a rising generation in science.[4]

Himmler, having given his personal assessment of 'decency' to Heisenberg now wrote to the physicist himself:

> . . . I am glad that I can now inform you that I do not approve of the attack in *Das Schwarze Korps* and that I have taken measures against any further attack against you.
>
> I hope that I shall see you in Berlin in the Autumn, in November or December, so that we may talk things over thoroughly man to man.
>
> With friendly greetings,
Heil Hitler!
Your,
H. Himmler.
>
> P.S. I consider it, however, best if in the future you make a distinction for your audience between the results of scientific research and the personal and political attitude of the scientists involved.[5]

Baron Carl-Friedrich von Weizsäcker was another brilliant young theoretical physicist who, like Heisenberg was fascinated by the advances which the last few years had given to the understanding of the structure of the atom. And like Heisenberg he had had an upbringing which had produced in him a highly developed sense of nationalism. His ties with the Nazi régime were, however, much closer than those of Heisenberg. His father was a party member, State Secretary and Head of the Foreign Office, so that it was natural that Weizsäcker, at the age of 21, should think of joining the party. Surprisingly, in the son's recollection, it was his father who persuaded him not to take this step. Although the father had the reputation of being as eager as Hitler to realise Germany's nationalist aims, according to the son he was secretly wary of his Führer's policies and in family circles confessed to believing Hitler to be something of a madman. Young Weizsäcker was not convinced and was impressed by the promise which Hitler held for the future of Germany, but he took his father's advice and stayed out of the party. But even though the young physicist admired Naziism in its early days, like others he was unhappy about the racial policies which went along with the party's creed.

The effect of Hitler's purges on the Jews was to cause more than a mental unease to some Aryan scientists. For some of them it meant a serious hindrance to their work since their close collaborators were Jewish. Otto Hahn, the physical chemist, had little sympathy with the Hitler régime and both he and his wife had good Jewish friends. From the earliest days he saw that the loss of Jewish scientists could well mean the beginning of the decay of German science. He had first-hand experience. Along with Fritz Strassman he was working in a small team with an Austrian woman who was also an excellent physical chemist and a Jewess: Lise Meitner.

In the middle years of the 1930's Hahn, Meitner and Strassman were looking into some of the problems of atomic structure which were exciting a good deal of interest in a number of other laboratories in Europe and America. They were following up some work of Enrico Fermi and were bombarding the element uranium with neutrons. Fermi had suspected that uranium, the heaviest known element in nature, which has an atomic number 92, might be converted by neutron bombardment into heavier new elements with atomic numbers 93, 94 or greater. The work was repeated in Hahn's laboratory and at first general agreement was reached with Fermi's conclusions, although Lise Meitner was reluctant to believe all that the evidence suggested. It was at this stage that Hahn felt the first real effects of anti-Semitism on his work. In July of 1938 Lise Meitner had to flee from Germany leaving Strassman and him to carry on with the work they had been doing together for four years.

The two Germans continued with their experiments which at the time of Lise Meitner's departure had reached a crucial stage. They again examined the products of the bombardment of uranium with neutrons and at first thought that they had found artificial radium atoms. Chemically, however, this is a very unlikely result and they took the experiment a stage further. On this occasion they found what they believed to be the element barium among the products of the bombardment. To Hahn and Strassman barium was as unlikely a result as was radium, and so they rather disbelievingly repeated the experiments yet again. The result was the same: Barium, in infinitesimally small quantities, but, nevertheless, barium. The reason why the result of the experiment was so surprising was that barium is a medium-sized element with an atomic number about half that of uranium. There was no precedent in chemistry to suggest that either barium or any other medium-sized element could be obtained by breaking down a heavy element such as uranium.

Hahn was convinced that the result which he and Strassman had got was not a case of mistaken identity. Faithful to his Jewish friend, Lise Meitner, Hahn wrote to her in Sweden, where she had now settled, and told her his latest news. It was Christmas of 1938 and Otto Frisch, Lise Meitner's nephew, had travelled over from Copenhagen where he was working, to spend the holidays with her. He arrived to find her poring over the letter from Hahn. Frisch listened to the news she had for him. His reaction was precisely that of Hahn's after the first experiment: he did not believe it. On that particular day Frisch was much more keen to talk about an experiment of his own. He had put on a pair of skis and his aunt had decided to walk on foot with him. It was under these peculiar circumstances – the by no means young lady physicist plodding behind the skis of her energetic nephew – that Frisch at last listened to what was being said. Lise Meitner kept on insisting that Hahn was too good a chemist to write a letter with such facts in it unless the facts were well-founded. Then and there the two began to discuss by what possible mechanism a uranium nucleus could give rise to a barium nucleus. The only solution which seemed at all feasible to them was one in which the uranium nucleus was elongated so that it formed a neck before finally being torn apart. As soon as Frisch got back to Copenhagen after the holiday he asked a biologist the name of the process by which bacteria divide themselves. The answer was 'fission', and Frisch adapted the word for the uranium process.

To Hahn his new discovery was no more than a piece of highly skilled chemistry of which he was exceptionally proud and it occurred neither to him nor to Strassman that it was a result which might have far-reaching implications. It was only after the publication of Joliot's work in Paris, which showed that when uranium is bombarded by neutrons and split, more neutrons are created so making a chain reaction possible, that the German atomic workers caught on to the full consequences of their work. The stored energy in uranium could be turned into a vast bomb. The possibilities, which had perhaps not escaped the American, British and French workers, were now freely discussed in Berlin, Göttingen and Münich and in laboratories in cities throughout the rest of Germany. Hahn himself, whilst well aware of the possibility of a nuclear explosion, was unwilling to believe that such a thing as an atomic bomb was practicable. Nevertheless one of his assistants, Flügge, published an article in the journal *Naturwissenschaften* in the summer of 1939 in which he discussed the mechanism of the chain reaction with neutrons and the possibilities of an explosion. It was Flügge who used for the first time the description 'Uranium machine' for the apparatus the German atomic physicists were to spend the next five years trying to build.

The State was not entirely unaware of the potential of its pure scientists. On 24th April 1939, the physicist Paul Harteck had written a letter to the War Ministry which is most sinister in its similarity to that which Einstein was to address to Roosevelt three months later. Harteck wrote:

> We take the liberty of calling to your attention the newest development in nuclear physics which in our opinion will perhaps make it possible to produce an explosive which is many orders of magnitude more effective than the present one . . . it is obvious that if

89

the possibility of energy production outlined above can be realised, which certainly is within the realm of possibility, that country which first makes use of it, has an unsurpassable advantage over the others.[6]

German atomic physicists could now be considered as a useful ally for the State. Among physicists there were one or two who were devoted followers of Hitler who were later to criticise Hahn, Strassman and Flügge for the irresponsibility they had shown for the interests of the Fatherland by publishing the results of their work. But these were still early days and Germany of all countries was surprisingly under-equipped for any rapid advances in practical atomic studies. There were only two adequately set-up laboratories and neither of them was government financed. They were sponsored by the Kaiser Wilhelm Gesellschaft, and were the Kaiser Wilhelm Institutes at Heidelberg and at Berlin-Dahlem. As for the hardware of physics which would be essential for a serious attack on the problem of producing a 'Uranium machine', there was little in Germany to give any laboratory a head start. Whereas in the United States there were several cyclotrons in operation, Germany's first at Heidelberg was not begun until 1938. But there was sufficient confidence for several scientists to consider drawing the attention of men in ministerial positions to the potential of uranium fission.

The letter to President Roosevelt which Einstein signed on 2nd August 1939 was not only a warning of the frightening power of an atomic bomb and a recommendation that the President should ensure a safe supply of uranium for the United States, it was also a warning that Hitler was probably on the point of taking similar advice via the son of a Secretary of State. The assumption of Einstein's letter concerning the channel of information from the younger von Weizsäcker to Hitler was wrong, but the assumption that the matter was being considered at ministerial level was right. Already, in fact four months before this letter to Roosevelt, six physicists were called by the Ministry of Education and pledged to secrecy at a meeting at which the uranium problem was discussed. The physicists were Joos, Hanle, Geiger, Mattacuh, Bothe, and Hoffmann; they were the first members of a group later to become known as 'the Uranium Club'.

Fear of an atomic weapon was a ball bouncing in and out of Germany. It rebounded yet again in the autumn of 1939 when news reached Berlin that United States military authorities had begun to allocate funds for atomic research. But this was not the only remaining call necessary to rally German physicists to a common cause. This ball had in the first place only begun to roll as a fearful thing because the discoveries which originated it coincided with the heightening of political tensions in Europe. News of the American military atomic work coincided with the outbreak of war with Britain. There were sceptics about the ultimate feasibility of an atomic weapon on both sides, but the paramount importance of being first with it, if it could indeed be built, was as obvious to German scientists as it was to their American and British counterparts.

When war came in September 1939, Germany had made a start with two of the most significant weapons of the 20th century: the atomic bomb and the rocket. In the case of the rocket, ten years of work had produced concrete results in the form of working models; within weeks of the opening of hostilities, two rockets were safely recovered after highly successful flights. Knowledge of uranium fission, however, was scarcely 10 months old by the time Dornberger's group at Peenemünde had totalled 10 years' experience. There was little enough to suggest at that time that the combination of these two weapons, the rocket-borne nuclear bomb, in a further 10 years' time would be the dominant factor in world politics.

Even before the outbreak of war a considerable number of first-rank German scientists had been called to arms. The Peenemünde project in particular was badly hit by the call-up. By the spring of 1940 Hitler had struck the rocket group from his priority list and many of its best men were being lost to the army. Some attempts were made to draw the attention of the German Propaganda Ministry to the seriousness of the situation by suggesting that Britain had learnt her lesson by the irreparable loss of Moseley in the First World War and that her scientists were being exempted from military service. Dornberger at Peenemünde took the matter into his own hands and unknown to Hitler managed to get 4,000 technically qualified men withdrawn from the armed forces and put at his disposal. Many physicists found themselves in uniform, but the most useful to the uranium project were returned to their trade quickly enough. The young von Weizsäcker, for example, was back busying himself with atomic physics as early as the first few weeks of September 1939.

Not all German scientists were willing to make contributions which would further Hitler's progress. During the war some physicists, among whom the best known was the distinguished Max von Laue, carried on their research with the intention of making no addition to Germany's war effort. But others who had no sympathy whatever with Hitler's methods, such as Otto Hahn, and again those who had unvoiced doubts and mixed emotions, such as von Weizsäcker and Heisenberg, became involved to different degrees in the new project.

In 1939 coordination between the various factions of German physics was embarrassingly bad. Soon after the outbreak of war the 'Uranium Club' discovered that it was not the only group interested in the application of uranium fission to the manufacture of a weapon. The discovery was made when it found itself competing for supplies of the precious uranium with a group which had similar aims. The group was sponsored by none other than the Army Ordnance Department – which was also responsible for the rocket research work. Nor were these the only interested parties. Yet another group, which was to have the Postal Ministry as its unlikely source of income, was working under the industrialist von Ardenne. But the atomic physicists were not alone in discovering that their efforts were being weakened by competition rather than being strengthened by cooperation. Dornberger too was later to discover that his was not the only group to be considering rocket development and that unknown to him he had competitors ready to take the field. One night at a dinner party after a meeting with Hitler, Pleiger, who was responsible for the Reich's coal production, let it be known that he was in charge of Germany's rocket development. Pleiger was completely unaware of Peenemünde's existence. Only following a long correspondence in which he actually tried to employ some of Dornberger's staff did Dornberger break the news to him.

The embarrassing situation in which the Uranium Club and Army Ordnance found themselves chasing the same supplies of uranium was settled by an agreement to pool information. For its main research centre for the project the War Office had chosen the Physics Institute of the Kaiser Wilhelm Gesellschaft in

Berlin-Dahlem. But even the choice of this Institute was something of an embarrassment for the physicists who had been nominated to carry out the uranium work. Its director, Peter Debye, a formidable scientist who had won the Nobel Prize for chemistry in 1936, was a Dutchman. When Debye had first been offered the appointment he had taken the precaution of not following the condition of assuming German citizenship which was normally required for the acceptance of a professorship at a German university. He had gone to the extent of getting special permission from his Queen to keep his Dutch citizenship and a written assurance from the German Foreign Minister that acceptance of the post would not automatically require German citizenship. Debye had therefore to be removed with as little fuss as possible. Soon after the War Office made its decision to take over the Kaiser Wilhelm Institute he was visited by an official and told that although he could now no longer enter his laboratory he was at liberty to stay at home, have whatever money he needed and, if he felt he could profitably spend his time that way, write a book. Instead, Debye accepted an invitation to lecture in the United States. For him at least the timing was fortunate. During the tour Germany declared war on Holland and so provided Cornell University with a new professor of chemistry.

With Debye's departure the atomic physicists could now begin to map out the theoretical lines they believed they would have to follow in order to exploit nuclear energy. They decided to concentrate on two problems; they knew the problems were difficult, but if they could be solved they would be the most likely to give fruitful results. Atoms of the same element, but with differing atomic weights (or 'mass numbers') are known as isotopes. The properties of different isotopes of the same element are, as might be expected, very similar, and isotopes can usually be separated from each other only with great technical difficulty. 993 of 1,000 naturally occurring uranium atoms have mass number 238, the remaining 7 have mass number 235. The first problem then was that of the separation of pure $U235$, which might be used to produce either controlled nuclear energy or an explosive bomb. The second was that of using ordinary unpurified uranium, along with a substance which would slow down neutrons and so make them capable of splitting the $U235$ of the isotopic mixture. This system again should theoretically be capable of producing a controlled nuclear reaction. It had the advantage of not requiring a solution to the immense problem of separating and purifying the uranium isotopes.

Within a few weeks of their first conferences the physicists had made encouraging calculations. By December of 1939 Heisenberg was considering the possibility of using an atomic pile consisting of layers of uranium and of either heavy water or pure carbon as a moderator to slow down neutrons for a controlled nuclear reaction.

But the more deeply the theoretical aspects were examined the more obvious it became that to put theory into practice would be a huge undertaking and would require the application of full-scale technical resources. There was a limit to what could be learnt from models and small-scale experiments. By the autumn of 1940 a group at Berlin-Dahlem under Wirtz had built its first model pile: it produced no energy but it did point the way to more experiments using more uranium and more heavy water, both of which were in short supply and which required expensive engineering on a large scale if these substances were to be produced in quantity and quality. Through all the brief number of years of their attempts to produce nuclear energy this lack of adequate facilities was to be the most serious hindrance to the progress of the atomic workers. During the summer of 1940 von Weizsäcker saw that a uranium pile should be capable of producing not only the medium-sized elements which are the fission products of uranium, but also the heavy isotope U239, among other products. In Germany at the time no one was to know how crucial this piece of theory might have been. In America the possibility was quickly realised. A transformation product of U239 was found to be the element plutonium, Pu239, and Niels Bohr and John Wheeler had shown that it could be broken up by slow neutrons. Since the Allied atomic workers were able to get plutonium from the relatively plentiful U238, and since it was fairly easily purified it could be, and indeed was, used in an atomic bomb. But the existence of plutonium was never realised by the Germans even though von Weizsäcker had written a report predicting the formation of a transuranic element which might have advantages over U235 in a bomb.

Great importance was now being attached to the production of heavy water. After the invasion of Norway every attempt was made to step up the production of the Norsk Hydro Company where heavy water was being made at the rate of about 10–20 litres a month. A number of physicists including Harteck, Jensen and Wirtz set about devising methods of pushing up the output and within a few months they increased it tenfold. It was during 1941 that the German workers considered that they had made their most important progress. By the middle of the summer Heisenberg had managed to build a model pile at Leipzig using pure uranium and a heavy-water moderator and the results convinced him that if only he could enlarge this model he would be able to produce atomic energy.

This was the period of optimism and confidence in their own abilities. The war was going well and for Hitler victory had followed victory. For the scientists victory over their particular problem seemed, if not just round the corner, then not too far distant. What form their ultimate victory would take was uncertain. If Heisenberg could improve his Leipzig pile and produce atomic energy this would be a triumph of physics, but would it be anything else? Would it lead to a super-bomb, for example? It might. Certainly the possibility was discussed eagerly enough among German physicists, just as were the advantages of Germany and Hitler owning a weapon which according to theory had potentially frightening powers of devastation. Even before this mood of optimism had set in, the physicists who had no liking for the Hitlerian régime in which they were living discussed with one another their misgivings; to a few the role of Sorcerer's Apprentice with a broomstick about to get out of control at any minute was not a happy one. Von Weizsäcker recalls discussing with Otto Hahn the possibility that their work might lead to a bomb. For Hahn, who had discovered fission, it was a disturbing subject and in a high state of emotion he said: 'If my work would lead to Hitler having an atomic bomb I would kill myself.'

In September of 1941 von Weizsäcker again discussed the implications of an atomic bomb, this time with a physicist working with Manfred von Ardenne's group financed by the Postal Ministry. He was Fritz Houtermans: a half-Jew whose career as a physicist had been chequered to say the least. Working in Russia in the 1930's he had been imprisoned and beaten up for suspected espionage. Repatriated in early 1940 he found himself in scarcely better conditions in a Gestapo gaol. Eventually he was released to work under von Ardenne. Houtermans was well aware of the progress which the rest of the uranium project had made during the early part of 1941 and when he heard that Heisenberg and von Weizsäcker were planning a visit to Copenhagen and to the greatly respected Niels Bohr he contacted Weizsäcker. Houtermans is quite clear in his recollection of the line the discussion took. Like the rest of the atomic physicists he was sufficiently impressed by the progress of the work that summer to believe that an atomic bomb was not too far away in effort and time. Precisely how much effort and time was still necessary to turn theory into reality no one knew: no one that is among scientists, and they alone could plot the destiny of the still unborn weapon. What seemed important to Houtermans was that they could make the decision to prevent its birth, and so prevent a monster falling to the use of Hitler – or to anybody else for that matter. He therefore suggested to von Weizsäcker that this visit he and Heisenberg were to make to Copenhagen would be an opportune time for them to take Niels Bohr into their confidence, tell him of the existence of the Uranium Club, and get him to give what Houtermans calls 'general absolution' guaranteeing that no atomic bomb would emerge from the German effort.

Undoubtedly Houtermans was not the only physicist to consider that Niels Bohr was the one man whose influence among scientists was sufficiently great to unite them in a common cause: that of restricting their work for the good of humanity. Bohr was to Houtermans and others the Pope of Physics, who in some way could grant this strange 'general absolution'. But there was another reason why Bohr might be an ideal sounding board for German physicists with problems of conscience. Bohr was still living and working in his own country: German-occupied Denmark, and was therefore accessible. It was well known that he was respected and even revered as much by Allied scientists as by Germans; moreover, it was suspected that he was in contact with the Allies.

For Heisenberg the circumstances of this visit in the Autumn of 1941 cannot have been pleasant. Copenhagen was a city occupied by German troops and hostility to the conquerors' intellectual representatives, if not so overt as in any other invaded European capitals, was inevitable. The Germans had made the usual initial attempts to improve the relations between themselves and the subjugated citizens; an Institute was set up to try to further German and Danish cultural relationships, but no one from Bohr's laboratory put in an appearance there. Any meeting with Germans which might be interpreted as collaboration was frowned on whether it was attendance at a Cultural Institute or informal talks with a visiting scientist. Therefore when Heisenberg appeared in Copenhagen, although his Danish colleagues were well aware of his strong sense of loyalty to them and believed him to be genuinely concerned for the plight in which they now found themselves in the occupied city, they were most chary in their approaches.

Why was Heisenberg in Copenhagen? According to Houtermans' conversation with Heisenberg's travelling companion, von Weizsäcker, it was in an attempt to suggest the possibility that German physicists would be prepared if necessary to guarantee that no atomic bomb would come out of their present efforts. Houtermans himself, however, had not talked directly with Heisenberg, although he had tried to do so. Heisenberg's own interpretation of what he thought he could achieve by a discussion with Bohr is considerably less specific than Houtermans' proposal. His method of reasoning against atomic bomb production is complicatedly mixed with his strong sense of nationalism. In his recollections to the author, Heisenberg said that he and his fellow physicists

> . . . could argue in two ways. We could say, one can make atomic bombs, therefore one should try and make atomic bombs. We could also argue by saying, well, it requires such an enormous industrial effort that it will actually weaken our war effort during the next years and probably the bombs won't be ready before the end of the war, therefore it's no use to make atomic bombs. Now, since in this way, the physicists seem to have a strong influence on the further development, I felt that it was good to ask for his advice. At the same time I felt that Bohr and I were very good friends and he knew my attitudes to problems of this kind from earlier years, so I felt that still this work would be a good basis for confidence and for conversation on this problem.[7]

In other words, there was a possibility that the war might end before the atomic bomb could be built in Germany, and therefore Heisenberg felt that he might with a good conscience consider preventing its completion since the cost and industrial involvement of such a bomb would slow down Germany's war effort.

Such a course of action would be loyal to both mankind and to Germany.

During Heisenberg and von Weizsäcker's visit there were several meetings with Danish scientists and Bohr entertained the pair at his home. Whatever impression the Germans intended to leave with their hosts they succeeded only in heightening the Danes' awareness of the German involvement in atomic weaponry. To friends, Bohr's recollection of conversations were that both Heisenberg and von Weizsäcker were convinced that Germany was now assured of an ultimate victory and that if they did not believe this, then Danish physicists were simply not facing up to realities. There was nothing in the bearing of these Germans to suggest to the Danes that their attitudes were a pretence to disguise some finer motive for the visit.

It is easy to believe that Heisenberg went to Copenhagen feeling that he was in a position to argue from strength. Work in the atomic field had gone so well in 1941 that even when the war was over he was to consider it the most important year of progress for the uranium project. And in the field of war there was the same pervading mood of optimism. The Tripartite pact between Germany, Italy and Japan stood firm, Rommel had had sweeping victories in the western desert and German forces were within striking distance of Moscow. These were the days before the Russian winter offensives, before Pearl Harbour, and before the entry of the United States into the war. There was therefore every reason for the Danes to be embarrassed by the visit of German physicists and every reason for Niels Bohr to be extremely cautious during his conversations with Heisenberg.

Eventually Heisenberg did manage to arrange to meet his old teacher alone and to talk to him where they could not be overheard. Heisenberg made it plain that he had something of importance which had been troubling him for many months which he wanted to discuss. It is significant, however, that Heisenberg and Bohr differ widely in their recollections of *where* they met. According to Heisenberg it was in a dark street not far from Bohr's home – where Heisenberg could be quite sure there would be no eavesdroppers. Bohr recalls it having taken place in his own office at the laboratory. It appears that the conclusions from the talk were as indefinite in the minds of the participants as was its location. Heisenberg, in his own recollection, was trying to do two things: he was trying to tell Bohr that he now most certainly believed that atomic bombs could be made, and he was trying to discover what were Bohr's 'attitudes to problems of this kind'. According to Heisenberg his news was

> . . . a kind of a shock for Bohr because Bohr realised from my question that I believed that atomic bombs could be possible during the war. Now this was such a shock to Bohr that his mind was concentrated on this problem and so in some ways he probably did not realise the other side of my question, and so he replied rather vaguely that of course in the war the Governments would always ask the physicists to work for them. . . . But I could see from his face that he was very worried about the problem.[8]

However, Bohr approached the conversation in a hostile frame of mind which Heisenberg could not have realised. It was an acutely delicate situation for him to meet Heisenberg in the conditions either of them describe since Bohr disapproved strongly of any form of collaboration with the enemy. That Heisenberg was his friend and one-time pupil made no difference to the fact that Copenhagen was occupied by German troops. He understood from Heisenberg that some problem connected with atomic fission had been worrying him for the past two years and that a discussion might help solve it. He suspected that Heisenberg had some sort of confession to make. Bohr therefore agreed to discussions along these lines with the intention of taking nothing more than a passive role, and the resolution to give no information whatever to Heisenberg which might be of any possible use to the German war effort. It is undoubtedly true that he was shocked by Heisenberg's belief that an atomic bomb might be decisive in the conclusion of the war if the war lasted long enough. At that time Bohr did not know of any of the advances that had been made towards a fission weapon. But Heisenberg's assertions that the worried look on Bohr's face (no matter whether he saw it in a darkened street or in a well-lit office) was due to the intense concern the information caused him, and that Bohr was so troubled that he failed to grasp the innuendoes of the rest of the conversation, are unlikely conclusions. Bohr was merely looking troubled at the awkward line the conversation was taking and his only concern at that moment was to give nothing away.

That the conversationalists cannot agree on where this potentially vital discussion took place points to the fact that Heisenberg's cautious attempts to reveal what was on his mind never
crystallised into unambiguous, easily recalled sentences. One thing
is certain: Bohr at no time got the impression that Heisenberg
wanted to discuss the possibility of his slowing down the German
uranium project, nor that he wanted to discuss a pact among
scientists to prevent the manufacture of an atomic bomb. The
Pope of Physics was not asked for nor did he give any kind of
'general absolution'.

Heisenberg returned to Germany no happier, and most certainly no wiser, after his interview with Bohr. Bohr was left in a considerably unhappier frame of mind, and wiser to the extent that he
now believed as a result of his conversations with Heisenberg that
the Germans were deeply involved in the uranium problem. It
was news he would pass on to the Allies as soon as he was able.
Heisenberg, on the other hand, had little idea what either the
Americans or the British were up to: but he was fully convinced
of the possibility of ultimate success of his own project. By early
1942, only a few weeks after the Copenhagen visit, he had carried
out the experiments with pure uranium metal which were so
successful that he was to write, 'Here then was the proof that the
technical utilisation of atomic energy was possible . . .'.
But this was the pinnacle of German progress towards an
atomic bomb. Atomic energy was possible, but it had still not
been achieved. And even if it could be achieved then its utilisation
in the German war machine would now require a huge application of resources on a scale which would make the expenditure up
to 1942 seem trifling. The members of the Uranium Club realised
that to commit their country to bringing their work to its fruitful
conclusion – an atomic bomb – would require technology to
be marshalled in a fashion completely unfamiliar to them and involving money, men and materials in quantities which no scientist
had yet had cause to dream of. It was this realisation – or admission – which caused the turning point in the German uranium
project.
In order to draw the attention of top government officials to
what they believed to be their most successful year's work, the
German scientists decided to hold a meeting at which they could

report their results. In addition to the chief sponsor of the meeting, Bernard Rust, the Minister of Education, it was intended that the most influential of Hitler's aides should be present. Formal invitations were sent to the highest party officials, including Himmler, Keitel, Goering and Bormann. But nuclear research was apparently not of sufficiently pressing importance for any of these to attend. Nevertheless, on 26th February 1942 the meeting was held at the Reich's Research Council in Berlin where a number of directors of war research listened to Heisenberg, Hahn and other physicists lecturing on the utilisation of atomic energy.

After this meeting administrational changes began to take place in the organisation of the uranium project. Its control now passed to Abraham Essau, an ardent Nazi, who had lectured at the Research Council Meeting, and on 6th June 1942 the physicists were again reporting their results: this time to Albert Speer, Minister of War Production. It was at this second meeting that the physicists told of their hopes and their fears. Their hopes were that eventually energy would be got from a uranium pile with little expense and effort. And their hopes were that soon, though how soon they could not guess, uranium isotopes, from which they could make an atomic bomb, could be separated. Their fears were that the effort involved in separating these isotopes might technically bankrupt Germany. Besides this, Heisenberg was well aware that, as a result of the war reaching a critical stage, Hitler had ordered that only the technical projects which would come to fruition within six months, and which could be of use in the war, should be financed. The physicists could give no encouragement to this sort of short-term thinking; when asked how long it would take to build an atomic bomb Heisenberg had replied: three years. It is hardly surprising that, following this meeting, Speer made the decision not to invest in a huge atomic programme. The nuclear physicists' work would continue, but on the same small experimental scale of the last three years. Whatever chance there had been of Germany making an atomic bomb had now slipped by.

The sky over northern Germany on 3rd October 1942 is recorded as having been clear and blue. It was a day to please the Peenemünde engineers. If the thing were to be a success, at least they would be able to see some of the spectacle and not have to watch it disappear into a bank of sea fret after a few seconds.

Peenemünde had changed. When von Braun had visited it on his Baltic coast holiday a few years before it had been a lonely strip fringed by pine forests and the sea. Now it was a complex of camouflaged laboratory blocks, liquid-fuel generating plants, airstrips, hangars, launching pads and conspicuous chimneys. On that October day, von Braun, now technical chief of the complex, was preparing to oversee the climax of 10 years' work. His administrative chief, Dornberger, had taken up a position on top of the Measurement House where he could see von Braun's propulsion engineers carrying out last-minute work before countdown. This was to be the test launch of a vehicle of the A4 rocket series.

The tension of the count-down was heightened by a relay of the count over loudspeakers. But no artificial aid was necessary to emphasise the gravity of the passing seconds. The danger was real – there had already been deaths and injuries in the series of tests; and the price of failure was high – only a few weeks before, a rocket known as the V3 had plunged into the Baltic 9,500 yards from the gantry after its motor had failed.

At zero of count-down the flow of oxygen vapour from the stern ventilators of the A4 stopped. Ignition was perfect. An orange flame followed the shower of sparks pouring from the deflector of the launch ramp. The great metal tube of the rocket rose with what was to the observers apparently paradoxical slowness: in reality only four and a half seconds passed before its nose cone tilted to the east and its speed began to increase.

Still in view to the watching technicians, the A4 passed through the sound barrier: the first vehicle to travel at supersonic speed. It was success. And the rocket was still travelling. At first there was some confusion and argument among the engineers as to whether it was only partial success; a jagged vapour trail had appeared from the exhaust as the speed mounted to 2,000 m.p.h. There need have been no fears: the A4 was merely passing at a previously unmentionable speed through stratified air-currents. It disappeared from view as a small white metallic reflection of the sun over the Bay of Swinemünde: it was moving at 3,000 m.p.h.

The engineers, the technicians and the military men flung themselves into each other's arms, shouting and laughing.

A reconnoitring Messerschmidt was later to bring back the news that the rocket had travelled 196 miles at a speed of 3,300 m.p.h. and had deviated only 2½ miles. It would be enough to take a bearing of St Paul's Cathedral from the French coast and be assured of a direct hit on London.

At a celebration party held at Peenemünde that evening Dornberger gave a little speech. He said:

> This 3rd day of October, 1942, is the first of a new era in transportation, that of space travel . . . so long as the war lasts, our most urgent task can only be the rapid perfection of the rocket as a weapon. The development of possibilities we cannot yet envisage will be a peacetime task.[9]

If the scientists and engineers of Peenemünde had, in the flushed moments of success, forgotten the immediate purpose of their work, then Dornberger had not. The necessity of his reminder, as he later discovered, was well founded.

In February of that same year the members of the Uranium Club had not thought it politic to suggest to the directors of war research that a colossal effort be poured into the atomic project. By June they had been content to see Speer, the Minister of War Production, put their scheme near the bottom of his priority list. Dornberger was as aware as Heisenberg that Hitler was anxious to finance only those schemes which would come to maturity and be of use in the war effort within six months. But Dornberger was one of the powerful advocates of the rocket group and the atomic group possessed no one of his conviction. Not only would he rush into corridors of power where angels feared to tread, but he could carry with him an intense belief in the inevitable success of the work being carried out under his administration. Wherever he felt he could get money for his scheme he would ask for it; throughout 1942 his main effort and purpose was to have Peenemünde put at the top of Speer's priority list, no matter what the consequences. Moreover, he jealously preserved the titular guardianship of his ten-year-old technological baby. More than once he had to battle off take-over bids for control, as for example when private industry, seeing the possibility of development and profits, tried to have Peenemünde turned into a private limited company.

Consistently he fought and consistently he won. And his advocacy was aided by one of equal, if perhaps differently orientated, persuasion to his own: that of Wernher von Braun.

At the time when the Uranium Club members thought they saw success ahead but were not prepared to recommend that Speer should commit any of his resources to the effort, the rocket group had still not managed to launch a completely successful vehicle. If Dornberger and von Braun were pessimistic then they did not attempt to put their reservations into words. Occasionally, however, disturbed undercurrents were spotted by those who could look down from high enough places. A Gestapo report of 29th June 1943 read:

> In January, 1943, there was a serious disagreement about the continuation of the Peenemünde Project. Involved were, among others, General Keitel and General Dornberger.
> The Führer called in all participants after each one was first obliged to state his point of view in writing. All military men were shown out within one or two minutes, because they were unable to answer the decisive questions of the Führer. Dr von Braun was the only one who talked for 30 minutes and was able to answer the precise questions of the Führer tersely and clearly. The Führer decided, therefore, according to the proposals of Dr von Braun.
> It is said, in addition, that the Führer has not yet a clear enough knowledge of the daily little wars between officers and engineers in Peenemünde. . . .[10]

On 7th July 1943 Dornberger, von Braun and the Peenemünde instrument engineer, Steinhoff, were ordered to fly to Hitler's headquarters to report their progress. The Führer, tired and hollow-eyed, sat between Speer and Keitel to watch the Peenemünde group's film to which the climax was the A4 tests over the Baltic. The film was edited to show the grandiose scale of the schemes. Its main purpose was to impress Hitler. It showed the huge structures of the rocket test frames, the gigantic assemblages of scaffolding and all the genuinely arresting power of the successful A4 launching. Hitler was suitably impressed. Only he dare break the silence which followed the screening. It was a silence of approval. He asked immediately for more technical details. Dornberger recalled in his memoirs a pronouncement from Hitler during that meeting that, 'Europe and the world will be too small from now on to contain a war. With such weapons humanity will be unable to endure it. . . .' Before the discussion broke up Hitler took von Braun aside for a private discussion. As he did so Dornberger reminded Albert Speer of a promise he had once made of a professorship for von Braun. When Hitler took his leave from the group he congratulated von Braun on his promotion.

Dornberger had at last got his top priority, von Braun had his professorship, and Hitler now believed that he had within reach his annihilatory weapon.

But how much were the men who were providing the means aware of the ends which their Führer had in view? They were, to use von Braun's metaphor, happily milking the golden cow; but could they see, and did they mind, who might get kicked in the process?

In a note to the author von Braun wrote:

> With the tight press censorship imposed by Hitler, the abuses of his régime were not nearly as visible to the average German as they were to an outsider who had free access to the international news media. For this reason, I must say, more by way of a statement than as an apology, that I never realised the depth of the abyss of Hitler's régime until very late and particularly after the war, when all those terrible abuses were first published. I guess until about a year before the war's end I shared the feelings of most Germans that while Hitler was unquestionably an aggressor and a conqueror, that this put him more in a class with Napoleon than with the devil incarnate. While right from its beginning I deeply deplored the war and the misery and suffering it spread all over the world, I found myself caught in a maelstrom in which I simply felt that, like it or not, it was my duty to work for my country at war.[11]

The producer of the Peenemünde film had closed it dramatically with a single sentence projected on to the screen: 'We made it after all!' It is possible that Hitler misread this statement as an assessment of the rocket as a weapon rather than a self-congratulatory description of the A4 test. Even so he had not put all his eggs in one basket. The rival to the A4 was the Luftwaffe's Fi-103 – the V1. This was a small jet-driven winged vehicle with a wing-span of about 25 feet. It was cheaper than the A4 (which was later to become known as the V2) though by comparison it had certain disadvantages. Army Ordnance had known nothing of the initial stages of the development of the Fi-103, though when von Braun heard of it he looked on it as healthy competition to the Peenemünde project. He thought the new pilotless plane well-founded from every viewpoint, but had his reservations when comparisons with the rocket were drawn. For example, although it would carry nearly the same explosive charge as the A4, the lower impact speed of 420 m.p.h. of the Fi-103 would give it a smaller destructive effect; the fact that, unlike the A4, it required fixed launching ramps suggested that it could be destroyed easily from the air, and its low speed and flight path would make it vulnerable to fighter and anti-aircraft attack. But notwithstanding such criticisms, the Luftwaffe also got its top priority for the V1.

Ironically it was a V1 which gave away the Peenemünde game to British Intelligence. In May 1943 an aerial reconnaissance interpreter in London scanned a photograph of Peenemünde and saw on it a pilotless plane: a V1. By August 1943 several hundred RAF bombers had delivered 2,000 tons of high explosives and incendiaries in a single raid on the coastal area. The whole complex was left almost devastated; the development laboratories and the administrative buildings were churned up; only the pre-production works, which were heavily protected by concrete casing, escaped extensive damage. The research workers' living quarters were particularly heavily stricken. 735 personnel were killed including one of Dornberger's leading scientists, Thiel, and his whole family. This was the beginning of the end for Peenemünde, though its progeny was to have its revenge on parts of Europe before the end finally came.

The A4 succumbed to a change of title: V2 or Vergeltungswaffe 2 – the revenge weapon. On the night of 5th September 1944, the first offensive rocket was put to work. It was launched towards Paris from a site south of Liège. Three days later the official offensive on London began. The effectiveness of the revenge weapon was evaluated in Germany by a study of obituary notices from local London newspapers. Addresses were matched with a street plan and when there was a cluster of deaths in neighbouring homes the result was assumed to be the impact effect of a V2. As soon as British security became aware of this casual but conclusive statistical analysis newspaper editors were prevented from including addresses in obituary notices.

As their report on the Peenemünde project of June 1943

showed, the Gestapo were well aware of the divisions of opinions between the project engineers and its military men. Himmler was apparently well informed of whatever schisms there were in the community. Unknown to Dornberger, he summoned von Braun to a secret meeting at his headquarters and put to him the hypothesis that research work under the S.S. might be provided with far greater scope than under Army Ordnance, with its insecure hold on priority funds. Von Braun refused to be drawn into the intrigue, but intrigue nevertheless overtook him. On 15th March 1944, Dornberger was called into conference with Field Marshal Keitel at Hitler's headquarters at Berchtesgaden. There he was told that on orders from Himmler, von Braun and two engineers, Riedel and Gröttrup, had been arrested by the Gestapo to face charges in an S.S. court. They had been reported as saying that they never had any intention of making a weapon of war from their rocket and that they had succumbed to Dornberger's pressure for no better reason than that this was the only way to get money for their experiments and to confirm their theories. The object of their work had been space travel. Von Braun was suspected of opposing the use of V2's against Britain and of planning to escape to England in a small plane taking his rocket secrets with him.

Dornberger's line of argument to secure the release of his chief collaborators was that of the interests of the rocket programme: they were indispensable for its completion. And if Dornberger had any cause to suspect their loyalties then there were others who were equally suspicious of his. While he was advocating the release of the trio of engineers to one of Himmler's S.S. Security officers he was warned that there was a fat file of evidence which if necessary could be taken down from its shelf and used in evidence against him should the A4's development be delayed. These Kafkaesque methods of security were by no means peculiar to the Gestapo. American secret agents played a similar game. At the same time as Dornberger was hearing of his well-documented records, an analogous file was in preparation in America, to be used if necessary against the scientific head of the U.S. atomic bomb programme, Robert Oppenheimer. In the German's case the political evidence against him was never used; in the American's case, it was.

Eventually von Braun, Riedel and Gröttrup were released. Later in the year, when Dornberger was considering resigning, after S.S. General Kammler had been appointed General Commissioner for the A4 programme above his head, it was von Braun and Steinhoff who tried to persuade *him* not to abandon *them*. They argued that irrespective of personal views, wishes and ambitions, the A4 should be developed, perfected and delivered fit for action; ultimately, they believed, rockets would be used for space exploration. Dornberger did not submit his request to be relieved of his duties.

Even by the middle of 1944 supreme German authority was demanding mass production of the A4. Its designers knew that to carry out these orders would result in a missile far below the standards of perfection for which they had hoped. Compromises had to be made in the engineering of a rocket which production lines could be capable of turning out in quantity. Nevertheless, in spite of the breakdowns which were a result of the panic demands, in the first month of operations 350 rockets were launched. Between 5th September 1944 and 27th March 1945, 5,500 V2's were fired from the front; of these 2,000 dropped on London and 1,600 on Antwerp. During the whole of this time the Allied armies were advancing on Germany from the west.

News of the V-weapons projects had been leaking from Germany during most of the war period. Hints had come from the usual sources; captured soldiers, escaped prisoners and partisans had carried sufficient information to make the Allies nervous of the outcome of Peenemünde from the start. But the same sources were not giving Allied military and scientific Intelligence any really reliable reports about the German uranium project. In America both pre-war and current issues of German science journals were minutely examined along with local German newspapers to try to find some clue as to where the principal nuclear workers might be posted. There had been more specific clues. A Berlin scientist had passed on the word through Norwegian partisans that the most important German work had been moved from Berlin to a safer place. The most probable location of this laboratory was given by a letter from a prisoner of war which U.S. censorship had discovered. It referred to nothing more pointed than a 'research laboratory numbered "D"' in an envelope postmarked Hechingen. It was conveniently coupled with a report that Heisenberg was living near this town.

It is perhaps a tribute to the past achievement of German science that when reconnaissance planes brought back photographs of unidentifiable large-scale installations on the French coast, in the absence of any more conclusive evidence, it was suggested that these might be bases for uranium piles, or even plants for producing radioactive poisons. Such was the confidence of Allied Intelligence in Germany's physicists that when the invasion of France began, in June 1944, troops were equipped with detectors which would locate radioactive materials.

The detectors were never put to good use. The June 1942 meeting which the German physicists had had with their Minister of War Production had put paid to any possibility of priority technological effort in the nuclear field. There were no sprouting atomic plants for the invasion troops to stumble across. Since 1942 research had continued only under university auspices. But in spite of this there was no let-up in the intensity of the work, for all that with each passing month of war it became more and more likely that the results of the effort would never be more than academic.

Heisenberg had been made director at the Kaiser Wilhelm Institute of Physics where, in order to house the large uranium piles which were planned, an underground laboratory had been built. But progress at the Berlin laboratory and at those at Leipzig and Heidelberg was always to be seriously hampered. German industry by the summer of 1944 was overloaded. War production was putting it under such a severe strain that projects which did not have the Minister's top priority could no longer rely on supplies. Heisenberg's planned experiments had to be postponed repeatedly since industry was producing uranium in such small quantities. There were also fears for the supply of heavy water. Germany's chief source of this precious liquid was the Norwegian Hydro-Electric Company's plant. In 1943 this was put out of action by an air attack. But before this stolen spring ran dry two tons of heavy water had passed between Norway and Germany; this, according to the physicists' calculations, was just enough to build a working atomic pile. The pile was never built; throughout the winter the Allied air attacks on the German industrial centres took their toll. In the early months of 1944 the supply of uranium slugs came to a temporary halt after the raids on Frankfurt. This in no way deterred the experimenters. They continued to calculate, and to make models from their calculations. Shortages and unreliability of deliveries of raw materials were now not their only worries. The conditions under which they had to experiment were becoming intolerable. At Berlin–Dahlem bombing had forced the main work underground: a model pile was built in the shelter, and experiments were carried out on it during some of Berlin's heaviest raids. With stoical understatement Heisenberg reported that the work was 'naturally to some extent hindered by the raids'.

On 15th February 1944, the Kaiser Wilhelm Institute of Chemistry received a direct hit in an air-raid. Still the work went on. By this time the Institute of Physics had moved to a village retreat: Haigerloch. There Walther Gerlach, the chief coordinator of nuclear research, had found a spot which might more adequately protect his physicists from Allied bombs. In a cellar cut into solid rock he equipped a laboratory where the uranium pile could be rebuilt.

Even though German research and development were now living on borrowed time there were still parties deeply interested in a possible outcome of the uranium research. Heisenberg had no intention of providing the false hope that it might be the country's saviour. When Dornberger's rocket group approached him to ask whether they might be able to look to atomic energy as a means of rocket propulsion he gave no promises of any description. But at the same time he was not willing to believe that the Allies' scientists might have moved faster than Germany's. In July 1944 Goering sent one of his A.D.C.'s to visit Heisenberg with the news that Intelligence had collected information through Switzerland which said that an atomic bomb was being prepared to be dropped on Dresden in two weeks. Goering wished to know what the chances were that the Allies had such a weapon. Heisenberg replied that the probability was very small. He heard nothing more.

The Allies were of course no less interested in German progress towards an atomic bomb than was Goering in theirs. Already the project in America had reached such a stage that some scientists and a very few politicians were only too well aware that if this weapon really could be built before the war's end then its owners might well pocket victory with horrific consequences. There was talk of Hitler's revenge weapon. Was it to be revenge by atomic bomb?

The United States War Department had equipped a mission known as 'Alsos' to follow the advancing troops into Italy to see what signs of nuclear research could be found. There were no surprises and few significant discoveries at the Italian universities. In the odd words of its chief instigator, General Leslie Groves '. . . the first Alsos mission was most successful. Indeed, its accomplishments so far exceeded what we had considered possible that its conclusions were generally discounted.'

With the invasion of France the Alsos Mission was expanded and packed off with a greater urgency and fear. Its leader was Colonel Boris T. Pash. Its scientific head, Sam Goudsmit, was a well-known atomic physicist who had spent the early years of the war working on radar and knew little of the American progress towards an atomic bomb. The choice of a scientist who knew in advance none of the secrets he might stumble across on his mission was not a piece of military mismanagement. Goudsmit was chosen as a competent physicist who, if captured, could give away nothing of value.

Goudsmit was in London preparing to set out across Europe with Alsos when the first V1 raids on the city began. The little information he had, told him that the raiders might be equipped with nuclear warheads. It was a worrying time for the half-briefed scientist. At that stage of the war air-raid sirens were being only scantily acknowledged by Londoners, and during his time spent in England Goudsmit had become as indifferent to air-raid shelters as the rest of the population. But now when the sirens wailed, Goudsmit had to take himself off to half-empty shelters with his incommunicable, but real enough, fears.

In Goudsmit's mind there was no doubt that, if a German uranium project existed at all, then the distinguished physicist who would be most deeply implicated in it would be Werner Heisenberg.

The Alsos mission travelled through France behind the advancing armies picking up scraps of information wherever it could. At Strasbourg Goudsmit carefully sifted the apparently innocuous papers left behind in the office of von Weizsäcker, who by this time had fled. Letters addressed to 'Lieber Werner' and memoranda which talked of 'large scale' experiments, and of a 'special metal', which could be nothing other than uranium, confirmed his suspicions. This information, that Germany's most able physicists were concentrating all their experimental and theoretical know-how on the uranium problem might have caused the Alsos leaders to recommend necessarily drastic action by the Allies. But von Weizsäcker's office had other news to tell. In a group of papers were computations which the scientific members of the mission were soon able to see applied to the theory of a uranium pile. There was no doubt from these calculations that, compared with the work going on in America at that moment, the German progress had been insignificant. They had not separated any pure uranium-235. It was inconceivable that a German atomic bomb could be on the point of completion in some Berlin bunker or Bavarian redoubt.

Nevertheless, with invading armies sweeping in from both east and west, still the work in the bunkers and the redoubts did not stop. As late as February 1945, a group under Wirtz at Haigerloch had gathered together almost sufficient material to build a uranium pile in the cellar cut out of rock. More uranium and more heavy water were needed. But these were in Berlin – a besieged city now under such constraint that there was reliable transport neither in nor out of it. The man holding these supplies in the capital was Walter Gerlach, the physicist who for the past year had been given overall charge of uranium research. A remarkable statement made after the collapse of Germany to Sam Goudsmit on the Alsos Mission tells graphically of the turmoil in the lives and the minds of the atomic scientists during the death throes of the Reich. The statement was given by a scientific publisher, Paul Rosbaud – an Austrian citizen. Gerlach looked on Rosbaud as a personal friend and let slip many indiscretions which the head of a nuclear research programme might properly have been expected to keep to himself. During the February 1945 period as the Red Army advanced, and as a new breach in the German lines was expected daily, Berlin was in panic. And it was in Berlin that the agitated Gerlach sat with the vital and valuable supplies he knew were needed at Haigerloch.

Late one evening Rosbaud had a telephone call from Gerlach asking to see him since Gerlach intended to leave Berlin next day. With no thought whatever for security he added that he intended to take 'the heavy stuff' with him. Rosbaud asked, 'Where? To Werner [Heisenberg]?' Gerlach made no reply to this, but when Rosbaud asked, 'What will he do with it?' Gerlach answered, 'Perhaps business!'. It was an ill-considered and frightening remark.

Next day Rosbaud went to see Gerlach and stayed to talk to him for several hours. According to Rosbaud the conversation centred on their fears for the ultimate safety of the 'heavy stuff': about a ton of heavy water and a stock of uranium which were secretly kept in a shelter. Rosbaud made it plain that he believed the material should at any price be hidden away safely and kept for peaceful research after the war. Gerlach said, however, that he had asked for Heisenberg's word that it would be protected from attempts to destroy it. Rosbaud unhappily recalled Gerlach's ambiguous telephone remark of the previous evening, that Werner's intention was 'perhaps business'.

When next the pair met some weeks later Gerlach was again in an agitated mood and began the conversation immediately with the news, 'The machine's working'. What Gerlach had in fact just learnt was that the last measurements to be made on the uranium pile were in full agreement with theory. Rosbaud pointed out that the time taken to bring a scientific idea to technical perfection might be as much as 10 years, and that the Russians were already sitting on the doorstep. Gerlach, however, was excited at the prospect of the success of the nuclear machine since it might well be used by 'a wise government' as a bartering power for better peace conditions. This was perhaps yet another interpretation of the 'perhaps business' remark. He nevertheless had no illusions that such a government existed at that time in Germany. Rosbaud made every attempt to bring Gerlach and his flight of politico-scientific fancy back to earth. He replied: 'Suppose somebody would really be so stupid and try to enter into discussions with the other side, what do you think that I would do if I were [in] their place? Either kill all the physicists who ever have been engaged in all this work so that they can't do any more damage, or to [keep] them all for so long in a camp behind barbed wire till they have confessed everything they know about the machine or the bomb.'

There was a certain prescience in Rosbaud's tirade. Within a short time German physicists were to find themselves locked away, though without the discomfiture of barbed wire, until they told, albeit unwittingly, what little they knew.

Colonel Pash's Alsos Mission discovered the first uranium pile laboratory in the cellar of an old school house in Thüringen. In a pit near the middle of the cellar had obviously been planned a structure of uranium oxide blocks surrounded by heavy water. All the precious materials, however, had already been removed two days previously to the Bavarian hideaway on Gestapo orders. Pash organised his own pocket invasion force to take over Hechingen and the nearby village of Haigerloch, but it was a foregone conclusion that his Mission would only unearth there discoveries which would turn out to be something of an anti-climax. The wing of the local textile factory and the Haigerloch cave held all the tangible results of German wartime nuclear physics. But when Haigerloch was occupied on 22nd April 1945 there were prizes to be had. Among the scientists captured were von Weizsäcker, von Laue, Wirtz, Bagge, Korsching and Hahn. Werner Heisenberg, the physicist described by General Leslie Groves as being '. . . worth more to us than ten divisions of Germans', was not in the first bag. He had left Hechingen by bicycle a few days before Alsos arrived. He was captured a few days later. When asked by Sam Goudsmit whether, now that war was over and Naziism a thing of the past, he would consider working in America, he replied, 'No, I don't want to leave; Germany needs me.'

Just as Gerlach dashed from laboratory to laboratory in a futile attempt to organise some kind of success from his scattered resources, Dornberger too was trying to fan the dying embers of his rocket project into a final flame. As the frontiers of Germany shrank by the hour, he travelled from airfield to test station and to factory hoping to finalise the successful launchings of guided anti-aircraft rockets. He drove through war-scarred countryside on roads badly pock-marked by bombing attacks, passing as he did so airfields on which stood some of Germany's latest designs of fighters. These planes were unable to leave the ground: there was no petrol for their tanks. Yet the mass production of fighters in the cities' underground factories was still going on. The bitterest irony for Dornberger was that, had his most sweeping plans for rocket production been followed, there need have been no shortage of fuel to keep the A4's on the ground. Liquid fuel could have been manufactured inside Germany's borders in hydrogenation plants.

The bombardment of southern England and Antwerp went on until 27th March 1945, when the V-weapon corps was disbanded and turned into an infantry corps. For the military command to take this course of action was perhaps a justifiably desperate step; but to move the new infantrymen to the Hartz area to prevent a linking of the American and Russian armies, which was the next step, was optimism on too grand a scale.

The Red Army was by now approaching Peenemünde from the east; far to the south were American forces. But already the higher echelons of Peenemünde had taken the decision whether to surrender to the devil or to the deep blue sea. As early as January at a secret meeting senior staff members had almost unanimously elected to move rapidly south when the time came for decisions. And when the time did come, included in a series of panic conflicting orders from the ministry, from the S.S. and from local commanders were instructions to move south.

A rapid evacuation programme was put into operation to prevent the most precious and useful of more than 10 years' collected work falling into the hands of the Russians. Wagonloads of papers were taken to the Harz mountains to be hidden in mine galleries. On 6th April the executive staff, about 450 in all, accompanied by members of the Security Service, was evacuated to the Lower Alps near Oberammergau. The engineers who made this journey were never sure whether the presence of accompanying Security men was for their protection and to prevent them falling into enemy hands, or whether they were to be used as hostages in some form of peace negotiations. It is significant that the members of both the uranium research group and the rocket project group rated the consequences of their work and their own acquired knowledge to be on such a high plane, and they so mis-read the military situation in the Germany of 1945 that they seriously considered influencing an armistice agreement to give better terms for their country.

The last month of the war was a peaceful one for Dornberger and von Braun: they spent it collecting a suntan on the terrace of their quarters. They lay with the peaks of the Allgäu mountains above them discussing their future and the few decisions they would very soon have to make to influence it to the good. On 2nd March Dornberger, von Braun and a number of other leading members of the Peenemünde group left their little Alpine retreat for the American front-line. They surrendered to halted U.S. troops at Reutte.

Before the year was out 127 German rocket scientists had signed a one-year's contract to work in America with the U.S. Army Ordnance Corps. Others found their way to Russia.

The victors had valuable human booty in their hands. This group of men was the only one in the world which was capable of building operational rockets. The ultimate aim of many of its members had been above all else to achieve interplanetary flight; like other scientists and engineers before and since in similar situations in other countries, these men had allowed the dictates of war to further the means to the end in such a way that the first application of some of the most revolutionary research of the century had been as a weapon.

As many German wartime scientists are still aware, the stigma of being a scientific pioneer under Hitler's aegis is a special and persistent one. What can be said of it with the benefit of hindsight? How should these men be judged who carry it? How do they judge themselves? Wernher von Braun, a quarter of a century after the events, allows God alone a pronouncement:

> What is a man's duty who loves his native country, sees that it has fallen into evil hands, but also that its very survival is at stake? It is easy to answer this question from the safe shelter of a strong, well-governed nation, which is more or less at peace with the rest of the world. The answer does not come as readily when you live right in the midst of such a storm of events. Quite a few of my German friends and relatives who fought in World War II won some of the highest decorations in combat. Yet several of these same men were later executed for their part in the abortive attempt on Hitler's life, while the rest kept fighting the external war. All of them undoubtedly thought they were just doing their duty. I have come to the conclusion that it is better for a man to respect another man's conscience than to sit in glib judgment of him. The final judgment on matters like this is surely in better hands with the good Lord. I thought, and I still do, that under the circumstances, I did what was my duty.[12]

The Allies were afraid that the captured atomic scientists might fall into Russian hands, and so even though the Alsos Mission had deduced that these men had nothing of great value to tell, they were nevertheless moved well out of range of the front-line. To the Americans they were an embarrassment. The U.S. atomic bomb laboratories had made progress which was so incomparably greater than that of the Uranium Club that it was believed that the Germans could not possibly make any immediate significant contribution to the atomic work. Yet there was a lingering suspicion – it is still voiced today – that the Germans had secret achievements to their credit which Alsos had not unearthed. Therefore, as a final safeguard, the most important of the physicists were taken in custody to England and to Farm Hall, the Georgian house in the quiet Huntingdonshire village.

There were 10 in all: von Laue, Hahn, Heisenberg, von Weizsäcker, Korsching, Harteck, Gerlach, Bagge, Diebner and Wirtz. The life they lived behind the lightly guarded walls was easy and at first relaxing after the strain of the closing weeks of war in Germany. The custodians of Farm Hall, a British major, and mainly American staff, set the deliberately gentle atmosphere of the incarceration by allowing the prisoners to choose their own rooms. The gaolers were surprised to overhear that consciousness of rank among the scientists had not disappeared even in their present declassed conditions. Von Laue, the most senior, was allocated the best room. Its position led to the ruffling of the calm waters of the internment. Von Laue, who had a bladder complaint and who, like the others, was forbidden to leave his room by night, had from force of circumstances to empty his chamber pot out of the window. Beneath the window was an American guard who collected the waste on his head. Not unnaturally the wet soldier suspected that the elderly scientist had deliberately covered him in urine. The incident was smoothed-over by the senior military officers. Von Laue was asked to look prudently out of the window first in future.

The tenor of the place was kept such that the physicists would feel free to discuss whatever they had in mind. In particular it was hoped that they would discuss their nuclear research and its applications in Germany, since even the most casual of conversations was being recorded, unknown to the Germans, for examination by Intelligence. The details of the lectures which they gave to each other to pass the time of day, and the asides during the vaguely sporting activities which they got up as a relaxation, were all noted, examined in Britain, sent to the United States, and any reference which might disprove the Alsos Mission's assessment of the failure of the German atomic project searched for. The search was in vain.

Heisenberg, when the group was still in Europe waiting transportation to England, had already unwittingly assured his captors that the Germans genuinely had nothing to hide. It had been the first occasion that he and his colleagues had been able to gather together for many months and they were able to share their latest experimental and theoretical results. Heisenberg then suggested that the progress of their work was far more advanced than, unknown to him, Alsos' findings really indicated; the combination of their contributions deserved to be recognised and its scientific priority acknowledged. He was not to know that these advances were insignificant compared with what had been going on in America.

This was not the last over-assessment of their own potential which the Germans were to make. The conversations overheard at Farm Hall showed how they continued to err. Annoyed talk at being continually locked away for no apparently good reason was constantly heard. These physicists still strongly believed in the solidity of the international fraternity of scientists and they had the idea that if only some of their pre-war colleagues now at Cambridge could be told of their plight, then help of some sort might be forthcoming. Heisenberg even went so far as to suggest that deliberate steps were being taken to prevent any such contact with German science. He supposed that the British would argue:

> . . . if we tell Dirac or Blackett where they are, they will report it immediately to their Russian friend Kapitza, and Comrade Stalin will come and say: 'What about the Berlin University professors? They belong in Berlin.' And it is possible that the Big Three will decide it at Potsdam and then Churchill will come back and say: 'Off you go' and the whole group is to return to Berlin, and then we'll be in the soup. . . .[13]

Blackett later visited Farm Hall to discuss with the physicists how German science should be set on its feet again after the war.

The most disturbed period of the Germans' stay in England came when they were told that an atomic bomb had been dropped

on Hiroshima. Hahn was first to be told and, as later were some
of the others, was clearly horrified by the news. He, perhaps more
deeply than anyone, felt that his own discovery of nuclear fission
had led to a ghastly and unnecessary conclusion. The British major
gave him a little brandy to get over the shock. Already the conditions of the confinement had caused acrimonious exchanges and
bitter personal remarks to pass between these men. The news
brought out more bitterness; although they were genuinely perturbed by the news of the use of the bomb, its success clearly
demonstrated their own inadequacies. Hahn was too ready to rub
in the irony of the situation, as the bugged conversations revealed:

> *Hahn:* . . . If the Americans have a uranium bomb then you're all
> second-raters. Poor old Heisenberg.
> *Heisenberg:* Did they use the word uranium in connection with this
> atomic bomb?
> *Hahn:* No.
> *Heisenberg:* Then it's got nothing to do with atoms, but the
> equivalent of 20,000 tons of high explosive is terrific. . . . All I
> can suggest, is that some dilettante in America knows it has the
> equivalent of 20,000 tons of high explosive and in reality, it
> doesn't work at all.
> *Hahn:* At any rate Heisenberg, you're just second-raters, and you
> may as well pack up.
> *Heisenberg:* I quite agree. . . . I am willing to believe that it is a
> high pressure bomb and I don't believe that it has anything to do
> with uranium but that it is a chemical thing where they have
> enormously increased the whole explosion.[14]

The nine o'clock news told Heisenberg how wrong he was.
The group heard the official announcement of the dropping of the
first atomic bomb. That day this was not the only group of
scientists to discuss the morality of the use of the nuclear weapon.
There were similar gatherings handling the same awkward subject all over the world, but not in such disturbed conditions. Even
before the official announcement, the atomic strike had been
condemned by von Weizsäcker.

> *Wirtz:* I'm glad we didn't have it.
> *Weizsäcker:* I think it's dreadful of the Americans to have done it.
> I think it is madness on their part.
> *Heisenberg:* One can't say that. One could equally well say, 'That's
> the quickest way of ending the war.'
> *Hahn:* That's what consoles me.

After the official announcement von Weizsäcker offered a reason for the failure of the German effort:

> *Weizsäcker:* I believe the reason we didn't do it was because all the physicists didn't want to do it, on principles. If we had all wanted Germany to win the war we could have succeeded.
> *Hahn:* I don't believe that, but I am thankful we didn't succeed.[15]

It was von Weizsäcker who was most conscious of the wider moral and political implications of the use of the nuclear weapon. He had been well aware of these during the period that the German atomic effort looked most optimistic. And at the same time he was well aware that there were other problems which he had erased from, or allowed to sink below the level of, his consciousness. For example, after the war he was prepared to admit that whereas in the 1940's his conscience was attuned to the probable results of his scientific work he, along with his colleagues, was not exercising his conscience as a citizen; scientists conformed as a class, as did others living under Nazi rule, and chose not to ask what was happening to the Jews.

The younger members of the group, like Hahn, were by no means convinced by what they believed to be von Weizsäcker's rationalisation:

> *Bagge:* I think it is absurd for Weizsäcker to say he did not want the thing to succeed. That may be so in his case, but not for all of us.[16]

Could Germany have made a bomb if she had wished to? Two years after the war ended Heisenberg had no doubts about the answer to this question. He wrote:

> It could not have succeeded on technical grounds alone: for even in America, with its much greater resources in scientific men, technicians and industrial potential, and with an economy undisturbed by enemy action, the bomb was not ready until after the conclusion of the war with Germany. In particular, a German atomic bomb project could not have succeeded because of the military situation. . . . The immediate production of armaments could be robbed neither of personnel nor of raw materials, nor could the enormous plants required have been effectively protected against air attack.[17]

But here there is a certain *post facto* wisdom. In 1942 the German physicists knew neither the time it would take to produce an atomic bomb nor how long the war would last. All they knew for certain was that an enormous industrial effort would be required. But Heisenberg goes on to discuss what he regards as another most important fact:

> . . . the undertaking could not even be initiated against the psychological background of the men responsible for German war policy. These men expected an early decision of the War, even in 1942, and any major project which did not promise quick returns was specifically forbidden. To obtain the necessary support, the experts would have been obliged to promise early results, knowing that these promises could not be kept. Faced with this situation, the experts did not attempt to advocate with the supreme command a great industrial effort for the production of atomic bombs.[18]

But in a conversation overheard at Farm Hall Heisenberg had put this another way: 'We wouldn't have had the moral courage to recommend to the government in the spring of 1942, that they should employ 120,000 just for building the thing up.'[19]

Today the 'ifs' of the situation might seem irrelevant, but they are not uninteresting. If someone had had this 'moral courage', and if the war had lasted longer, then, in spite of the many protestations made to the contrary, Germany might have had the decisive atomic bomb in its armoury. It is true that Germany was far inferior to America in manpower, resources and materials, but yet she brought a phenomenal V-weapons project within sight of a decisive conclusion; had it not been carried out and seen to be believed, the same arguments could have been used to discount it ever having begun. The cost of the American atomic bomb was about £500,000,000 but by 1940, two years before the main Allied atomic rocket began, £40,000,000 had already been spent on the A4 rocket project, and this was only the start of the effort. Dornberger and von Braun had had the 'moral courage'. They had been powerful advocates for themselves. A remark of Korsching's at Farm Hall showed the climate under which German scientists had argued the value of their projects; the success of the atomic bomb showed, he said, 'that the Americans are capable of real cooperation on a tremendous scale. That would have been impossible in Germany. Each one said that the other was unimportant. . . .'[20]

When the German physicists rightly saw that the continuation of their nuclear work on a large scale would hinder the war effort they removed from their shoulders an enormous moral burden: that of deciding whether to make and use an atomic bomb. The burden was to rest in a singularly awkward fashion on the backs of the scientists of another nation. 'Thank God,' said Heisenberg 20 years later, 'that we couldn't build it.'

The chronicles of the death throes in war of a nation inevitably turn up for posterity examples of recriminations, self-pity and rancour among the men who failed in the fight. Episodes in the wartime career of German scientists might seem unedifying, but at least the participants can plead justifiable differences born in a nation running to defeat. But for one of the victors, Britain, the picture of science being harnessed to war has its embarrassing blemishes where personalities and their ideas have crashed into each other. Even today strains of the bitterness of conflicting views exist among those most intimately concerned with H. T. Tizard and F. A. Lindemann.

C. P. Snow has said of Sir Henry Tizard, 'Tizard's was the best scientific mind that in England has applied itself to war.' The claim is sweeping. The nature of the scientific mind is difficult to define, but no matter what criteria for comparisons are used, there can be little doubt that Lindemann's contributions to science were at least as great as those of Tizard. That Winston Churchill's view of the scientific mind applied to war did not coincide with that of C. P. Snow is seen from the fact that it was Lindemann rather than Tizard whom he appointed to be his personal scientific adviser. He was the first Prime Minister ever to take on a scientist in such an intimate relationship. Churchill wrote in *The Second World War*:

> . . . There were no doubt greater scientists than Frederick Lindemann, though his credentials and genius command respect. But he had two qualifications of vital consequence to me. First, as these pages have shown, he was my trusted friend and confidant of

twenty years. Together we had watched the advance and onset of world disaster. Together we had done our best to sound the alarm. And now we were in it, and I had the power to guide and arm our effort. How could I have the knowledge?

Here came the second of his qualities. Lindemann could decipher the signals from the experts on the far horizons and explain to me in lucid homely terms what the issues were. . . .[1]

Whatever the relative merits of their individual contributions, between them Henry Tizard and F. A. Lindemann (later Lord Cherwell) were to have greater influence on Britain's application of scientific invention to warfare than any other men with scientific training. They were, moreover, during the years when their influence on governmental policy was most effective, on such bad terms with one another that their rows were an embarrassment to the men with whom they worked. Professor Blackett describes how, 'On one occasion Lindemann became so fierce with Tizard that the secretaries had to be sent out of the committee room so as to keep the squabble as private as possible.'

Curiously, it was in Nernst's laboratory taking their doctorates in pre-1914 Berlin that these two who were later to involve themselves so deeply in war against Germany, and in embittered squabbles with each other, first met. Apart from their interest and ability in physical science these two men had little in common, and yet in their youth they became good friends and their long careers in science and administration were to have close parallels. Lindemann's family was rich and his background aristocratic. He was born at Baden-Baden in 1886, the son of an American mother and of an Alsatian father who had become a naturalised Englishman. He was a fluent German speaker, but the fact that he had been born on German soil was always an irritation to him since he never considered himself to be anything other than an essential Englishman. He was a man of strong prejudices who elected clearly defined groups of friends and enemies. After his death his character was to be frequently judged by different writers in the same uncompromising black and white terms in which he judged others.

Tizard had been born a year earlier than Lindemann; his name was as un-English as Lindemann's, but his family had had roots in the country for several centuries and his background, unlike Lindemann's, was both unworldly and thoroughgoing English.

Throughout his life he worried about money. He refused Lindemann's offer to share rooms during their student days in Berlin because he felt that Lindemann was so much better off than he. If Tizard's letters are used to judge, then he was a rather dry Englishman who had no difficulty in keeping his humour under control. What Tizard and Lindemann had in common were strong characters and powerful wills. By the time the characters eventually rubbed up against each other with the wills firmly opposed, Britain had passed through one war and was getting ready for another.

When the First World War came Tizard left a comfortable Fellowship at Oriel College, Oxford, and travelled to London with H. G. J. Moseley to join up. Moseley was commissioned in the Royal Engineers and went off to his death in the Dardanelles, and Tizard was commissioned in the Royal Artillery and went off to train Territorials. Before long, however, he had transferred to the Royal Flying Corps in order to be able to do experimental aeronautical research at Upavon. There he worked on the new equipment being introduced for the first time into aircraft: radios, cameras, bombsights, bombcarriers, and any other pieces of technology which could turn the transport machine (it was only six years since the first flight of the English Channel) into a scientific weapon of war.

Tizard learnt to fly and was soon taking trips to Farnborough to see his friend Lindemann who was carrying out research as a civilian at the Royal Aircraft Factory. Neither of the young experimenters was entirely satisfied with his status in wartime life and neither had much hesitation in letting the appropriate authorities have his views. For Tizard it was a hurt letter of complaint along with some of his fellow officers that, as a scientist responsible for developments from phosphorus bombs to searchlights, he was both under-rank and under-paid. Lindemann's dissatisfaction in life was of an entirely different nature. The unfortunate circumstance that his mother had happened to be taking the cure at Baden-Baden when giving him birth cast some doubt, along with his name, on his true nationality. Lindemann, therefore, had not been given a commission in the Army or Royal Flying Corps, and so in order to remove any ambiguity of suspicion of his position he wrote a firm letter asking for a commission and protesting both his aristocratic French ancestry and his indubitable Englishness. Neither letter produced positive response, but this did not prevent either Lindemann or Tizard from throwing himself with great eagerness into his work.

One piece of work which Lindemann in particular applied himself to was to investigate what happens when an aeroplane goes into a spin. In the early years of the war a plane which got out of control and began to spin almost invariably crashed and killed the pilot. There is considerable academic argument as to exactly what priority Lindemann can claim in having put planes into clockwise and counterclockwise spins. Priority aside, there is no question that Lindemann did pilot RE8's to great heights, throw them into spins, memorise readings of air speed, time, number of turns and other essential experimental flight data, push the stick forward in spite of the opposite instinctive reaction, and purely as a result of his own calculations, pull the machine out of the spin and back into stable flight. In later life Tizard acknowledged Lindemann's coolness in making accurate measurements under difficult conditions, claimed that the Germans had mastered the spin technique first, pointed out that he himself had put planes in and out of spins before Lindemann and concluded that Lindemann's courage had been overpraised.

It was some years after the Great War before the childlike acrimony was to develop between these two brave and astute men. Lindemann, with some help from Tizard, was elected to a professorship at Oxford and began to transform the Clarendon laboratory from the out-of-date, understaffed, under-equipped, under-financed institution which it had become under the previous professor and to give it the foundations of a modern science department. Tizard returned to his Oxford Fellowship but before long was considering the offer of a new and important post which had been made to him: that of Assistant Secretary to the Department of Scientific and Industrial Research which had been formed during the war. The job was that of coordinating research for the Armed Services. Tizard satisfied himself that the secretaryship would not mean that he would have to sever his scientific connections, worried himself about how he might suffer financially by taking the job, and then accepted. Tizard stayed in governmental scientific administration for nine years, widening his role and increasing his influence. In 1929, to the surprise of many of his colleagues – though again money and Tizard's worry over it were determining factors – he accepted the offer of Rectorship of the Imperial College of Science and Technology.

In a short span of time, which had included four years of World War, Lindemann and Tizard had reached what for some might have been pinnacles of their careers. In academic terms their responsibilities were large, and their jobs had a permanent look about them. But they were still young men. Lindemann was scarcely 33 when he was elected Professor at Oxford and Tizard was 44 when he took the Rectorship at Imperial College. And along with this relative youth they were the owners of some invaluable skills. It needed only a few more years of maturity before the combination of their practical experiences of war and their understanding and administrative experiences of science was to put them both in positions of great influence at a crucial period of European history.

In May, 1933, one scientist who had been born in Germany wrote a letter criticising German national characteristics to another scientist who had also been born in German territory. The writer was Einstein: a Jew. The recipient was Lindemann, who in his youth had a marked prejudice against Jews, but who was now helping Jews find academic posts in Oxford. Einstein wrote:

> . . . I think that the Nazis have got the whip-hand in Berlin. I am reliably informed that they are collecting war material and in particular aeroplanes in a great hurry. If they are given another year or two the world will have another fine experience at the hands of the Germans.[2]

Both Lindemann and Tizard, because of their first-hand experience of air power at Farnborough and at Upavon, were conscious of the vulnerability of British cities to bombing attacks. Lindemann, now under the wing of Churchill, wrote to *The Times* and his letter appeared on 8th August 1934 with the heading, *Science and Air Bombing*. He deplored the attitude that there can be no defence against air attack, and went on:

> . . . That there is at present no means of preventing hostile bombers from depositing their loads of explosives, incendiary materials, gases, or bacteria upon their objectives I believe to be true; that no method can be devised to safeguard great centres of population from such a fate appears to me to be profoundly improbable.

Lindemann never had any doubts that if society was threatened then science should be prepared to play its full part in its protection. In the same letter he said:

> ... It seems not too much to say that bombing aeroplanes in the hands of gangster Governments might jeopardise the whole future of our Western civilisation.

It was this fear of Hitler's gangsterism which later drove many scientists to work they might otherwise not have considered to be within their province.

Pressure on the British Government to take some sort of action increased from a number of quarters and early in 1935 the Committee for the Scientific Survey of Air Defence held its first meeting. Its chairman was Sir Henry Tizard and it later became known by the more common informal title, the Tizard Committee. It was this group of scientists which encouraged the birth of radar and so successfully exploited it. The defence of Britain had been put on a safe scientific footing and the nation was soon to be thankful.

After the public and private badgering and lobbying in the cause of air defence which he had carried out with Churchill, it can only have come as a surprise, if not a shock, to Lindemann to have to be told that, unknown to him, it had now been decided to set up an Air Ministry Committee under Tizard. This, however, was not where the matter rested. As a result, in part, of Churchill and Lindemann's dissatisfaction with air defence administration, and the apparent lack of power of the Tizard Committee, after more high-level lobbying a new committee was formed under the Air Minister. This committee, the Swinton Committee, was charged to deal with political as well as scientific problems of air defence. In the position of sub-committee to the Swinton Committee was the Tizard Committee. When Winston Churchill was invited on to the Swinton Committee it was inevitable that he should make a condition of his acceptance, 'that Professor Lindemann should at least be a member of the Technical Sub-Committee, because I depended upon his aid'. And so Lindemann, Churchill's scientific right hand, became a member of the Tizard Committee, and the parallel lines which had been the careers of two scientists prepared for a disastrous meeting.

From the moment that Lindemann walked into his first Tizard

Committee meeting the pair who had once been youthful scientific colleagues loaded their big guns, took ranging action, and fired on each other for all they were worth. In Tizard's words they became 'mortal enemies'. Their battlefield was the best method of scientific defence of Britain, and their ammunition was words. As his allies Tizard had such eminent scientists as Professor A. V. Hill and Professor P. M. S. Blackett for committee members. Lindemann was not without covering strength and even if his support troops were not numerous, they included W. S. Churchill.

The paroxysms in the committee room, during the antagonistic clashes of these two men, were still being looked on with involved passion by men who felt themselves close to the events or to the personalities, even 25 years after the bitterness was at its height. C. P. Snow, for example, leaves little doubt as to where his sympathies lie. Of Lindemann he says: 'The professional scientists did not take him seriously as a scientist, and dismissed him as a cranky society pet. Scientifically his name was worth little.'[3]

Snow's knowledge of the events is a result of his having trod the corridors of power near to the committee room: of Lindemann's behaviour there, he says:

> It may have been – there are some who were close to these events who have told me so – that all his judgments at these meetings were due to his hatred of Tizard, which had burst out as uncontrollably as love. That is, whatever Tizard wanted and supported, Lindemann would have felt unshakeably was certain to be wrong and would have opposed. The other view is that Lindemann's scientific, as well as his emotional, temperament came in: it was not only hatred for Tizard, it was also his habit of getting self-blindingly attached to his own gadgetry ideas that led him on.[4]

Of this sort of comment, made at Harvard University by Snow in 1960, Lindemann's biographer, Lord Birkenhead, says:

> It is perhaps unnecessary to comment at length on the propriety of a British scientist washing such dirty linen in a foreign University. Sir Charles Snow's account of the dispute between the two men resembles a Victorian melodrama in which virtue in the form of Tizard is triumphant and the villain Lindemann hissed off the stage.[5]

At one point Lindemann was indeed most successfully hissed

off the stage. In his early weeks on the Committee he was appalled at the slow pace at which experiments on air defence methods were being carried out and was not in the least shy about expressing his views both forcefully and unambiguously. One of his favourite pieces of technology, one of the 'gadgetry ideas' referred to by Snow, was the small explosive aerial mine which he believed might well be suspended, perhaps by parachute, in the path of approaching bombers. Another was a method of infra-red detection and location of aircraft. These, and other grounds, provided plenty of fertile soil in which acrimony could take root.

Tizard soon found reasons to complain of Lindemann's personal comments to Winston Churchill which had given rise to 'ill-founded criticisms'. In reply Lindemann had more to say of his lack of satisfaction with the rate of progress of the Committee – and relations went from bad to worse. Within a few months the skirmishes had developed into the bloodthirsty verbal battles that so embarrassed their colleagues.

Before the end of 1936 Lindemann had been temporarily moved into the wings. As a result of the resignations of Blackett and Hill and of Tizard himself in protest over Lindemann's attitude and actions, the Committee had had to be dissolved. It was refounded three months later; Tizard kept his position as Chairman, and no invitation to rejoin was offered to Lindemann.

Freed from Lindemann's vindictiveness, Tizard went on with his Committee to some of the most important administrative work in recent British history: that of turning radar into a workable defence weapon. Lindemann, in defeat, had to try to push his ideas in other quarters. But Lindemann was not off the stage for long. In 1939 it was Lindemann whom Churchill, as First Lord of the Admiralty, took for his personal statistical adviser and kept him in the same post when he became Prime Minister. And so Lindemann, as a result of a fortunate friendship with a man destined for power, had at last eclipsed Tizard and become the most influential scientist in the country and, at that time, even in the world.

The Lindemann–Tizard affair is an example of ambition and willpower having run amok over personal relationships. That the scientists involved with the Tizard Committee were alarmed by the spectacle of the squabbles of two of their colleagues whom they had taken as their leaders, and that they were embarrassed by the necessity to take sides, there can be no doubt. The result, however, was in no way a disaster for Britain: only personalities were scarred, and Lindemann and Tizard's relationships were to improve considerably. What is significant is that by the time these two highly intelligent, verifiably brave, and fiercely patriotic men reached positions of power in Britain unequalled by any other scientist, they themselves had ceased to be scientists.

By 1927 Tizard had already written his last scientific paper: a publication dealing with the properties of carbon dioxide at high temperatures. His new administrative post had meant the inevitable petering out of the opportunity to carry out practical work and to publish. This step out of pure scientific research into the application of science had been a deliberate one taken by Tizard as a result of his own assessment of his own ability. He later wrote that he had already convinced himself that he would never be outstanding as a pure scientist. It was not a rare decision, but if all research workers in science were to make the same honest assessment and, the appropriate huge proportion having reached the same conclusion as Tizard, take the same action, science as we know it would not be able to continue; research depends as much on its second-rank men as it does on its Einsteins and its Rutherfords.

Surprisingly, although Lindemann was still professor at the Clarendon, he too was becoming more and more remote from pure science. Unlike some heads of modern science departments he did not publish large numbers of papers in co-authorship with young experimentalists and colleagues; except for a few rare papers, like Tizard he had ceased to publish by the 1930's. He was now veering from the obvious path of the career which seemed open to him as one of Nernst's most brilliant and promising pupils; his main concerns had turned to University affairs and the development of his laboratory.

Both men had now become the leaders of a new and henceforth fundamentally important breed: the scientific administrator. One time scientists in their own right, their most important task from here on was to grow other men's flowers: to be the judges and critics of other scientists' work with particular reference to how best this could be applied to the art of war.

The work of other men, to which they had to put their minds

and which they attempted to apply, ranged over a wide field. For example, they had to consider and encourage the development of novel high-powered oscillators in order to perfect microwave radar as a war tool vital to the process of killing off German night fighters and surfaced submarines, and as a navigational device leading to the blasting of German towns. Lindemann forced on experiments with tinfoil strips dropped from aircraft in order to jam enemy radar. Tizard, in 1940, led a mission to the United States to make a crucial improvement in Anglo-American relations by handing over some of the most vital secrets of British science of which he had been the foster-father.

These two men were the manipulators of other men's creations in one of the century's most fruitful periods of scientific creativity. They coordinated the best achievements in scientific war work and, no longer scientists in the sense that we now use the word, acted as the corporate conscience of many men much further down the line who were nearer the source of original invention.

The attitudes of Lindemann and Tizard seem clear. Whatever personal differences there were, they arose only as a result of disagreement as to the quickest and best method of achieving victory on the battlefields of land, sea and air. Their aims were the same: the common enemy had to be crushed. Looked at objectively, they were cold men for enemies to have to face. To them, war was war and had to be won by any means possible. Tizard's biographer, Ronald Clark, found it an illusion

> . . . that Tizard himself shrank from the fact that in an industrial society war is indivisible – that he believed there was some point along the bicycle path from the factory to the home where the morals of killing civilians suddenly changed.[6]

During 1942 critical decisions had to be made about the most effective method of launching bomber offensives on Germany. To both Tizard and Lindemann the most effective method was one which, with the minimum loss of British bomber-crew lives, would most economically put the German industry and war machine out of action, whether by direct attack, or by attack on the civilian population which served the industry and the war machine. Lindemann, now created Lord Cherwell and a member of the War Cabinet, studied a statistical analysis of the results of the crippling of British cities by German air bombardment, and the misery and disorder which the blitz created for its victims, and came to the conclusion that British bombers could wipe out the houses of the majority of the inhabitants in no fewer than 58 German cities and towns. Tizard disagreed with the policy, but only because he did not believe in Lindemann's arithmetic. Blackett, now Director of Naval Operational Research at the Admiralty, had done his own sums and come to the conclusion that Lindemann's estimate of the effectiveness of such bombing offensives was six times too high. Lindemann, the Cabinet Minister and intimate of the Prime Minister, had his way which, he wrote, 'would break the spirit of the people'.

War, in all its horror, was war. This was the first scientifically operated war, and in it many scientifically trained minds on both sides of the firing line, believing in the singularity and the justness of the cause, pressed on to victory with whatever means there were at hand. Lindemann's or Tizard's methods of reasoning must of course be seen in the context of the time just as must those of say Heisenberg or von Braun. Lindemann was living and working in a bomb-blitzed London and had based his calculations of his recommended bombing policy on the sufferings of the inhabitants of Hull and Birmingham. Tizard had recommended in 1940 that, if the worst came to the worst, and German landing craft swept on to the British coast, then electric light bulbs filled with mustard gas might be spread on the beaches. If the use of poison gas in the 1914-18 war is taken for comparison, then any such desperate action would most certainly have provoked retaliation in kind; nevertheless, Tizard's suggestion was made when Britain was in the most grave danger.

Was there such a thing then as a limit under any conceivable circumstances: a line of action, a scientific experiment, an application of a scientific device, a tool of war at which men would baulk and say, 'This far, and no further'? In the 1939–45 war there was only one significant point at which a number of men, worthy of some consideration, paused to take stock because of the nature of the line they now had to cross. For the most part these men were not scientific administrators like Tizard and Lindemann. They were almost all scientists working directly in the field of physical research. It was because of their intimate knowledge of the technological masterpiece of a weapon they had created that they had to bear the responsibility of stopping to take stock and to ask what was going to be the consequence of using what they had made. The weapon was the atomic bomb.

R. V. Jones was a young physicist who saw the interaction of Tizard and Lindemann at close quarters. He had worked under Lindemann at the Clarendon and in 1935 had been drawn into the problems of air defence, which was Tizard's concern, because of his original work on infra-red techniques as a means of aircraft detection. During the war as an Intelligence Officer he had intimate contact with the antagonisms of the two men and was able to tread lightly between them. Yet in their attitudes to the atomic weapon he found them curiously united. Both had encouraged early work on the weapon and yet both had grave doubts about its ultimate feasibility. Even in early 1945 when work in the United States was nearing its crucial phase neither appeared to want to have to face its tremendous consequences. In one conversation with Lindemann, Jones pointed out how inferior a position Britain would hold in the applications of atomic power after the war as a result of the progress made solely in America. Lindemann's attitude was one of, there's many a slip 'twixt cup and lip and of, what fools the Americans would look if the whole effort proved to be a ghastly multi-million dollar failure. Tizard, as late as March 1945, asked Jones if he 'really thought it possible to release energy from the atomic nucleus and make an explosion'.

Both men lacked the full awareness which only those most directly concerned with the atomic research could have. The out-of-character mystical rationalisation of both was that God never intended that so much power should be put into the hands of men. There is a strange addendum to Tizard's familiarity with the progress of atomic research. In later years he told Jones that during the whole of the time that he was Chief Scientific Adviser to the Government, from 1947 to 1952, he was not allowed fully into atomic energy secrets.

9: Warning Noises

The effect of Nazi dogma on German science was disastrous. The men who found that they no longer could live in a country where physics was bad physics if it was Jewish physics had to find homes in countries with more receptive scientific climates. The escapes from Europe of scientific celebrities such as Einstein and Freud were accompanied by Nazi trumpetings of disloyalty and good-riddance from German universities, and welcoming cries from British and American universities. The welcoming cries were genuine and heartfelt, even though, in a Britain still suffering from the effects of the depression, the stream of refugees increased the competition for available university jobs. But for all that, there was still a compensating element of self-reward coupled with the welcome. Some of the physical science departments of Western universities were expanding considerably both their scope and their staff: they could skim off Europe's cream and could benefit themselves as well as the refugees who were in the cream. But the chief benefit for the host nations – and nobody could have recognised the full potential at the time – was to be felt in the coming war which would soon engage science and scientists on a previously unimaginable scale.

The loss, therefore, was doubly great for Germany and the gain doubly great for America and for Britain. Lindemann, concerned as he was for Jewish scholars, had been by no means slow to realise what possibilities the situation in Germany held for his plans to improve the Clarendon laboratory at Oxford. As early as 1933 he was touring Europe in his chauffeur-driven limousine encouraging scientists to make new homes in British universities and in particular to join his group working on low-temperature physics at the Clarendon. America too had its self-appointed talent scouts who were ready to look over Germany's rejected talent. Einstein, one of the prize pickings, had already spent some time at Lindemann's college but it was pacing the lawn of a quadrangle of this college, Christ Church, that Abraham Flexner, the American educationalist, managed to persuade Einstein to take up a permanent post at his new Institute for Advanced Study at Princeton. Before the outbreak of war in Europe men of the quality of Max Born, Franz Simon (later Sir Francis Simon), Otto Frisch, Rudolf Peierls and many others had found their way, if not always immediately to cool cloisters and green quadrangles, then to bed-sitters near red-brick universities. Many of these men are still living and working in Britain, the younger ones now holding leading professorships in universities where they have been responsible for the education of a new generation of mathematicians, physicists and chemists.

In America the story was the same. Emigrés such as Enrico Fermi, Hans Bethe, Edward Teller and others who left the countries of their birth in the 1930's and never returned, provided a powerful shot in the arm for the science faculties of the universities of their new land and today the United States is still able to draw on their talent and their teaching.

But this flow of Jewish refugees into receptive new universities had consequences which were much more immediately significant than the stiffening of the intellectual climates of Oxford or of Liverpool or Birmingham. Just as in America, Szilard, Wigner and Teller were taking steps with Einstein to try to interest the highest circles of United States Government in their thoughts, so too in Britain the émigrés made their presence felt outside academic circles. These men came from the centres of German research; they had been, and in many cases still were, close friends of some of Hitler's most distinguished chemists and physicists, and so they could be expected to bring with them a good deal of information which might be vital for a government preparing for a scientific war. Within a few months of the beginning of the war, Tizard put on an official footing a scheme whereby refugee Jewish scientists could be discreetly interviewed so that anything of significance they had to tell could be passed on to the British Government.

But just as they did in America these émigrés to Britain brought with them something else to throw into the melting pot of academic thoughts of the late 1930's. They brought with them an awareness and a fear. To those outside Germany the excesses of Hitler's dictatorship were spelled out in a number of newspaper articles and in books and, at a more popular level, and certainly more forcefully, by cinema newsreels which were at the peak of their powers as media of propaganda. There were graphic scenes of anti-Semitic scuffles in Bavarian market towns, of the distasteful humiliation of anonymous old men with beards, and of jack-booted, blond young men throwing armfuls of Jewish literature on to bonfires of books. But the images to observers at a distance can only have been dulled. To those who had suffered directly, who had seen their homes ransacked, who had experienced the humiliation and the pain, and who had recognised the portents and the reality of the concentration camps, the signs were clear and in harsh, sharp focus. The fear was of Naziism, and the aware-ness was that any device which might be offered by Germany's scientists and technologists for use in any capacity of war could, and would, be used by Hitler against Germany's enemies. Some Jewish scientists in Britain, unsure guests in a strange country, were reticent and kept their fears to themselves; others, particu-larly social scientists, spoke their fears too loudly and had their motives suspected by their new colleagues; others succeeded in heightening the growing uncomfortable awareness.

The work on uranium fission which had been going on in German laboratories was of course no secret in Britain; know-ledge of it had arrived via the normal scientific channels of publi-cation and by correspondence between one scientist and another, and before the outbreak of war the study of fission was under way in British universities. There can be no doubt that the recognition by a number of refugee and non-refugee scientists of the immense potential of uranium as a weapon was transmitted to Lindemann, who in turn transmitted it to the man who had adopted him as his personal adviser: Winston Churchill. But for the Govern-ment's advisers to decide which horses to back at this early stage in the scientific war was no easy task. A uranium weapon was only one of many of Hitler's potential secret weapons which sat in front of these advisers, and these ranged from one-man death rays to ultimate global weapons as suggested by both scientific luminaries and by crackpots. In the case of uranium fission many held considerable scepticism for the ultimate feasibility of turning what was still only a series of laboratory experiments into a practical super-explosive. Even at this stage it was clear that the cost and the effort would be immense and that the end product might turn out to be nothing more than a useful source of heat energy rather than a weapon. Both Tizard and Lindemann had expressed their doubts. Nevertheless, Tizard put into motion operations for buying up supplies of uranium.*

How were Britain's younger academic scientists reacting to the political rumblings from within Europe? Many, who had only the slenderest of memories of the First World War, were now realising that they would not be able to avoid making a decision whether or not to involve themselves in the preparations for a new conflict, even though it might be months or even years before the Second World War was officially opened.

In February of 1933 in the Oxford Union, the motion 'That this House will in no circumstances fight for its King and Country' had been carried by a vote of 275 to 153. The Union has traditionally been a stomping ground for men from the humanities, and in the 1930's the proportion of science students at the university was far less than it is today. How far, then, did this Union motion truly reflect the views of young scientists during that pre-war period?

When R. V. Jones, the young physicist working at the Clarendon, heard of the voting on the Union motion, he felt that he ought to reassure his parents about his own attitude. He told his father, once a regular Grenadier, that should war come then, although he might not be involved in the way that the son of a Grenadier might be expected to be involved, he most certainly intended to serve in some capacity.

In November 1937 he was writing a letter about the possibilities of forming a volunteer anti-aircraft detachment in Oxford. He had made it his business to sound out the local townsmen – mechanics from Morris Motors, instrument makers from the Pressed Steel works, members of the local rifle club – and had come to the conclusion that the pacifist elements in the University

* Britain's part in the development of nuclear weapons is told in detail in the U.K.A.E.A.'s official history *Britain and Atomic Energy, 1939–1945* by Margaret Gowing.

had had little effect on the citizens of Oxford. So insignificant did he judge the influence of Gown on Town to be that he confidently offered to raise 1,000 men to form a Territorial battalion. His letter, addressed to the Instructor in Searchlights of the 1st Anti-Aircraft Division, added,

> Some of the dons are naturally keen. Professor Lindemann, for example, who is head of this laboratory, has offered to send this proposal direct to Mr Hore Belisha, but I am submitting it in the first place to you so that it may have the chance of going through more normal channels.

The normal channels, in this case the War Office, were penetrated and Jones was assured that Whitehall would 'bear in mind' his 'helpful suggestions'.

It was scientifically, however, that he believed he could best contribute should war come. Already he had begun experimenting at the Clarendon on the development of thermopiles as a means of aircraft detection, and in September 1937 this infra-red technique had advanced sufficiently to be able to distinguish one aircraft from another. However, as a useful method it had limitations which caused it to be pushed into the side-lines by radar: it could not penetrate cloud and it was not capable of giving aircraft range. In March of 1938 the Tizard Committee recommended that the system be abandoned. Jones, who had turned down some tempting astronomical fellowships in order to work on defence applications of infra-red, was now faced with the prospect of winding-up his research and considering what he should do next. It was during this period of personal and political uncertainty that Hitler's troops walked into Austria on March 11th. Within a few days Jones was writing to D. R. Pye, the Director of Scientific Research at the Ministry of Aircraft Production, on the effect of the *Anschluss* on the Clarendon:

> . . . The events of the past week have made the research people here realise that the position is more serious than they had thought. Yesterday one of them asked me what he should do in the event of war: he wanted to do something active, and pointed out there was nobody to tell him what to do. In the past, most scientists have tended to be conscientious objectors; following this spontaneous move, I investigated the feelings of other members of the laboratory, and found that out of eighteen people questioned, only two were now conscientious objectors. Most of the remainder wanted to do scientific military research, while one or two of the more

pugnacious would prefer to take more vigorous measures. This change of outlook is largely due to recent events, and I have little doubt that the majority of scientists will volunteer when the time comes. Oxford has always tended to be more pacifist than other sections of the community, so that the above figures probably underestimate the enthusiasm of the rest of the scientific world.

The main reason for Jones' letter, however, was not merely to inform the Ministry of the state of mind of young Oxford scientists, but to suggest that now was the time to be planning to apply their talents:

> The point is this: if war were to break out tomorrow the scientific directorates of the services would find themselves overwhelmed by volunteers, and much valuable time would be wasted in finding out what posts they were best suited for, and the necessary – and as far as I know unforeseen – expansion would have to be effected.
> I am suggesting therefore that the research workers in the universities should be asked what they want to do, and to state their lines of specialisation, should they elect to join the scientific staffs during wartime. You would then know your prospective personnel, and could arrange your necessarily expanded programme accordingly. The men could then be informed where they were to be stationed, and laboratory accommodation arranged. They could start practically at the outbreak of war, and no time would be wasted.

Jones received an interim reply from the Ministry saying that Pye would probably reply more fully, but he never did. To many scientists what Jones was suggesting was obvious and Pye was undoubtedly familiar with these ideas from more official sources which perhaps had less parochial concern than Jones's. But by the time war began, still no corporate policy for the use of Britain's scientific manpower had been decided on. On more than one occasion the Royal Society had attempted to make contact at ministerial level in order that its Fellows could be officially on hand to give advice at top governmental level and to ensure that there would be a direct line of communication with the scientific community. A Central Register listing the qualifications of 7,000 scientists was eventually drawn up for the Government. However, it was not until October of 1940 that the Scientific Advisory Committee to the Cabinet was formed which included members of Britain's most distinguished scientific club, the Royal Society.

Only in the cause of one vitally important piece of research was Britain's scientific academic community properly coordinated and ready in good time to meet the war: the realisation early in the 1930's that finding some way of preventing the Luftwaffe sweeping over the coast was not a desirable luxury but a pre-eminent necessity had led to the birth of British war science's golden boy: radar. It had become a foster-child of the Tizard Committee.

As early as 1935 Robert Watson-Watt, working with a small group of scientists at Orfordness on the Suffolk coast, and using improvements in the microwave reflection technique which he had developed over several years, managed to detect aeroplanes at distances of up to fifteen miles. As the possibility of war turned into a probability, so the range and accuracy of radar improved; as it became clear that it would be a practicable and vital force in the air defence of Britain, so it became certain that the cream of Britain's electrical engineers and physicists would be needed to develop and maintain a chain of radar stations to guard, in the first instance, London and the Thames estuary. It so happened that the nearest supply of physicists to Watson-Watt's expanded establishment at Bawdsey was the Cavendish laboratory at Cambridge. And so during the few months before the outbreak of war Tizard spent a profitable time preparing the way into radar for the brightest young men of this, the leading physics laboratory in the country, as well as placing others from a number of universities.

The safeguard of keeping the progress of radar hidden from Germany was obvious; therefore, the groups of physicists who were taken on familiarisation tours of the East Coast radar stations in the pre-war months were all of them British nationals, and all of them sworn to secrecy. Non-British were excluded from the project. This act, designed to protect the golden secret, was peculiarly responsible, not only for concentrating a high proportion of Britain's most outstanding young academics on to one crucial piece of defensive technology, it was also responsible for channelling a disproportionately large number of foreign scientists in Britain to work on a much more momentous piece of aggressive technology.

At this indeterminate stage of the war, life for many of the refugee physicists in Britain had its awkward sides. Many of these scientists had no intentions whatever of taking British citizenship and planned to return to their homelands just as soon as they were able. Joseph Rotblat, for example, was a young physicist who in May 1939 had been given a year's research fellowship by the

Polish Government to work under Chadwick at Liverpool University. When war broke out Rotblat had to stay on in England and offered his services as a physicist for war work; these were politely refused.

There was nothing, however, to prevent the émigrés who felt sufficiently strongly about making contributions to the war effort to do so on their own initiative in the fields which they could contribute most and which they considered to be the most urgent. As it happened there was one field from which they were not entirely excluded because, in part, it was not considered to need such priority measures as did radar: this was the field of nuclear research. Already in Britain academic scientists had begun to look into the practical possibilities of a chain reaction in uranium. At Imperial College under Professor G. P. Thomson the main experimental work was being carried out, while at other universities, basic theoretical work and plans for the extraction of uranium metal were matters which were being given as much consideration as the more immediate tasks such as filling sand-bags and fire-watching would allow. These were the first unsure gropings towards an atomic bomb. It was inevitable that when war came, the attention given to research projects which had only a long term future should wane. Energy from atomic fission was one of these projects. Prospects in Britain by the spring of 1940 were at a low ebb. In the first place the most talented of young scientists were being drawn off, if not into the armed services, then into work such as radar and armaments research where there was an immediate crucial need for original minds. Secondly, the results of the earliest attempts in Britain to initiate a chain reaction in uranium oxide had failed and had not given much reason to hold out any hope for a practicable short-term solution. At this time the thoughts of most Britons were on defence; scientists, led by the public and private attitudes of senior establishment men such as Lindemann and Tizard, were no exception. And without radar Britain might not have survived to fight an aggressive war. But it was the refugee scientists and those who had experienced Hitlerism at close quarters who were perhaps most acutely conscious of the reasons why a defensive concept against Germany was totally inadequate. The reluctant émigré, Joseph Rotblat, was sharply aware of the published progress towards nuclear fission and he reasoned, knowing that Hitler's Aryan scientists were by no means inconsequential, that if an atomic bomb could be made in Germany, then the only consideration which could prevent Hitler from using it would be the fear of retaliation. At the time Rotblat knew nothing of the work already going on in other parts of Britain and so, his offer to do war work having been coolly refused, he began some work on neutron scattering on his own initiative. Rotblat told his ideas to his professor, Chadwick (the discoverer of the neutron), who was at this stage just being drawn himself into the atomic project, and was destined to play a significant part in it.

There were several other refugee physicists who, knowing that their abilities were not welcome on top-secret projects, had time to spare to think about the theoretical problems of the elusive chain reaction – and the menace of Hitler. Otto Frisch was one of these. It was Frisch who, with his aunt, Lise Meitner, had realised the mechanism by which a uranium nucleus undergoes fission. By the winter of 1939 he had found his way to Britain and was reviewing the progress of nuclear fission to date in the Annual Reports of the Chemical Society. Niels Bohr had by this time come to the conclusion that ordinary uranium would either not support a chain reaction at all, or only a slowish one which could not possibly cause the devastating explosion which had been so much talked about, and Frisch repeated Bohr's news in his publication. There must have been many who read Frisch's article who nodded agreement to it. The great Lord Rutherford after all had said only a few years earlier, and before the discovery of fission, '. . . we cannot control atomic energy to an extent which would be of any value commercially, and we are not likely ever to be able to do so'. Frisch was now merely echoing the phrases of the father-figure of Cambridge experimental physics: or so it seemed. But the one man who most doubted the correctness of this article was none other than Frisch himself. Having written and published it, he re-read it – and re-thought it, with a little more care and considerably more apprehension.

Bohr's arguments had been based on the assumption that the material used would be ordinary uranium. Tentatively Frisch now began to consider what might happen if it were possible to make a sphere of almost pure $U235$.

Frisch, at that time, was working in Birmingham and was living in the house of Professor Rudolf Peierls, a fellow émigré.

It was to Peierls, naturally enough, that Frisch took his thoughts; together they had often discussed the prospects of fission. Now they reconsidered Frisch's hesitant work and, more or less on the back of an envelope, they came to the frightening answer that only two or three pounds of U235 in a fast neutron reaction could make an explosion many hundreds of times more devastating than the biggest existing bomb. The ultimate conclusion which these two still quite young physicists had to reach as a result of their startlingly brief calculation might have been quite different had they been working in the country of their origin in a time of peace. They now found themselves in an industrial city of a country just starting out on what could not be anything other than a great and frightful war, and knowing that, what they themselves had worked out on a small bit of paper, could well have been worked out in the same simple way in some university town in the enemy country, Germany. The time and the place were such that the conclusion they had to reach was not that they had hit upon the way to release a vast amount of energy for the good and the glory of man: it was that they had in their hands the recipe for a powerful military weapon.

Provided there was no gross error in the calculation, its importance was unmistakable. They wrote up their thoughts into a few simple typed pages which they headed, 'On the Construction of a "Super-bomb"'. But for all its simplicity, and probably because of it, this Frisch–Peierls memorandum was a remarkable document. In clear language which even non-specialists would have little difficulty in understanding, Frisch and Peierls first set out the problem, gave a theoretical solution and went on to the practical details of construction of a super-weapon. Their first point was to show how previous workers had overlooked the possibility of the use of pure U235. They pointed out how, in a fast neutron reaction using this isotope, an enormous amount of nuclear energy could be released before the material had time to blow itself apart; they estimated that a 5 kilogram bomb would be equivalent to that of several thousand tons of dynamite. They then went on to describe how the sphere of pure uranium of critical size would have to be made in two parts which would have to be brought together by some trigger mechanism when an explosion was needed. Once the parts were united, the bomb would explode spontaneously in less than a second as a result of neutrons from cosmic rays starting off the chain reaction. They also recognised and went on to discuss the problems of preparing pure U235; they suggested the use of a thermal diffusion method with a suitable gaseous uranium compound. In this technique the lighter of the uranium isotopes concentrates near a hot surface and can be progressively enriched. The practical difficulties of the scheme were acknowledged in the memorandum. At the time Peierls thought of the total cost of isotope separation as being about the same order as that of a battleship – but worth the investment for all that; this turned out to be a gross underestimate of what an atomic bomb would eventually eat up in manpower and materials, but although there were errors and approximations in the short document, it showed an amazing comprehension of how a weapon could be built.

A brief recipe for the bomb was not all that the Frisch–Peierls memorandum offered: it went further beyond the explosion and considered its effects. Here in plain terms the two physicists spelled out what a bomb would be capable of doing.

> In addition to the destructive effect of the explosion itself, the whole material of the bomb would be transformed into a highly radioactive state. The energy radiated by these active substances will amount to about 20% of the energy liberated in the explosion, and the radiations would be fatal to living beings even a long time after the explosion. . . .
>
> . . . Any estimates of the effects of this radiation on human beings must be rather uncertain because it is difficult to tell what will happen to the radioactive material after the explosion. Most of it will probably be blown into the air and carried away by the wind. This cloud of radioactive material will kill everybody within a strip estimated to be several miles long. If it rained the danger would be even worse because active material would be carried down to the ground and stick to it, and persons entering the contaminated area would be subjected to dangerous radiations even after days. If 1% of the active material sticks to the débris in the vicinity of the explosion and if the débris is spread over an area of, say, a square mile, any person entering this area would be in serious danger, even several days after the explosion. . . .
>
> . . . Effective protection is hardly possible. Houses would offer protection only at the margins of the danger zone. Deep cellars or tunnels may be comparatively safe from the effects of radiation, provided air can be supplied from an uncontaminated area (some of the active substances would be noble gases which are not stopped by ordinary filters).

L

The irradiation is not felt until hours later when it may be too late. Therefore it would be very important to have an organisation which determines the exact extent of the danger area, by means of ionisation measurements, so that people can be warned from entering it.

In a second portion of the memorandum the two physicists made the point that they did not feel competent to discuss the strategic value of an atomic bomb, yet they had some most pertinent comments to make concerning its likely military use. They stated their belief that as a weapon it would be practically irresistible, and significantly that it '. . . could probably not be used without killing large numbers of civilians, and thus may make it unsuitable as a weapon for use by this country'.

Frisch and Peierls went on to acknowledge the uncertainties in their estimates; however, it is clear from this document that they had not only confidence in the calculations which the most recent research had enabled them to carry out, but also that they had a full realisation of the implications of these calculations.

Both the nature and the scale of the end product, if it succeeded in its task, would be totally and momentously different from those of any other weapon man had yet managed to concoct.

Both men were well aware of the moral problems involved in making such a weapon though they seldom discussed these with each other. At that moment in time the overriding consideration was that if such a weapon fell into the hands of Hitler then it would be a disaster for them and for the whole world. Frisch who not long before had been forced out of his university post in Hamburg had never the slightest question in his mind that as soon as he realised that an atomic bomb was possible then Britain must try to make it; Peierls concurred. What they elected to be their duty was to put their thoughts before the highest civilian authority they knew and whom they believed would understand. By March of 1940 their memorandum was on the desk of Sir Henry Tizard. It was not for five and a half years that their predictions were verified when the first atomic bomb showed itself to a world still totally unprepared for its vast powers.

The result of Frisch and Peierls' memorandum was that of a galvanising effect on British interest in an atomic bomb which, until the facts contained in this document were circulated in the right quarters, had shown signs of dying away in the face of more immediately urgent priorities. Undoubtedly other factors – other ideas and other fears being raised by other scientists in different parts of the world – also had their effect, but the contents of the two émigrés' memorandum could not be overlooked by the men who understood it in Britain.

It is remarkable how closely both the springs of thought and the action as a result of this thought were duplicated by similar groups of scientists both in Britain and in the United States. Unknown to Frisch and to Peierls, seven months before their memorandum reached Tizard, Einstein had addressed his letter to President Roosevelt and it was this that had led to the 'Advisory Committee on Uranium'. Like Frisch and Peierls the two chief authors of this letter, Szilard and Wigner, were European Jews who were acutely aware of the deeper implications of the advances being made daily in this fresh field of nuclear research. And just as Einstein's letter had warned of the necessity of early action in view of the fact that von Weizsäcker, the physicist son of the German Secretary of State, was probably aware of the possibilities of a uranium bomb, so too Frisch and Peierls recommended that the activities of the Münich physical chemist Clusius (the inventor of a method of isotope separation) should be looked into. Just as Einstein's letter had pointed out how a single atomic bomb could conceivably destroy a whole port and its surrounding territory, so the Frisch–Peierls memorandum suggested that it might be used against a naval base and would likely enough cause at the same time a great loss of civilian life. Einstein's letter had been an alert to Roosevelt and a recommendation that physicists be taken into the President's confidence to give him the benefit of their advice on latest developments. The Frisch–Peierls memorandum, aimed at the British Government's advisers, emphasised the urgency by giving a bald description of how it might be possible to set about making the weapon and gave the first warnings of the uncertain but far-reaching effects which the first bomb might have whenever it was used.

It was not long before the contents of the historic memorandum reached Professor G. P. Thomson who was coordinating the uranium experiments in Britain, and within weeks a governmental scientific sub-committee whose responsibility was the bomb was in being. It was called the M.A.U.D. Committee; the origin of these initials is itself a comment on scientists' own realisation of their involvement and their assessment of their potential high value in the new scientific war. Any information from any quarter which might have some value to atomic research was being turned upside down and inside out to extract whatever useful knowledge it might hold. Any news of the work of scientists in Germany or German occupied territory was subjected to especially cautious enquiry. Not only published data but also apparently casual remarks, messages and letters were re-examined to see whether they had any hidden message giving vital clues. For example, during this period of flutter and uncertainty a reference to an unknown man called 'D. Burns' in an otherwise apparently innocent communication could perceptibly raise the pulse-rate of a physicist who would normally be both unexcitable and uninterested in cross-word clues. *D* is the chemical symbol for deuterium, the heavy isotope of hydrogen. Was there a possibility that under certain conditions deuterium could cause a nuclear explosion? It was just such an unlikely reference as this that caused a cable from Niels Bohr in Denmark to John Cockcroft in England to be subjected to the minutest scrutiny. This telegram had in it what appeared to be an innocent message for 'Maud Ray, Kent'. In no time the cross-word *afficionados* had replaced the *y* by an *i* and had turned the anagram into 'radium taken'. Was it possible that Bohr was trying to draw the attention of British scientists to some newly discovered use of radium in nuclear research? There were other possible interpretations of the cable. Only after the war was it discovered that a friend of the Bohr family did indeed live in Kent and that her name was Maud Ray and that the cable had included a brief message for her. It was this lady who had the enduring distinction of having her first name used as the title of the British uranium committee. Whether it was good security policy to use the smokescreen M.A.U.D. for a group dealing with highly secret processes of fission of an element for which the chemical symbol is U, is quite another matter. Indeed, many people who knew of the Committee's existence were already convinced that it was concerned with 'Military Applications of Uranium Disintegration'.

Under its chairman, Professor G. P. Thomson, the Maud Committee pressed Britain on towards the beginning of a nuclear age. Work was started at the universities in Cambridge, Oxford, Birmingham and Liverpool as well as in industry. There had never been any doubt that if the operation was to be a success then it would have to be organised on a giant scale and accordingly that industry specifically created as the giant of Europe, I.C.I., was drawn in. There were other less spectacular sources of aid to be tapped. The French physicists, Halban and Kowarski, who had made distinguished contributions to knowledge of slow neutron reactions, escaped to England as France crumbled bringing with them a priceless supply of heavy water. The Maud Committee arranged for them to work on its behalf at Cambridge.

The Maud Committee's first meetings had been in the spring of 1940; in not much more than a year of fervent involvement in its task it was able to present its findings in the form of two documents. The first was its report, *The Use of Uranium for a Bomb,* and the second its report, *The Use of Uranium as a Source of Power.*

The second paragraph of the first report made quite clear its assessment of the year's findings. It began:

> We should like to emphasise at the beginning of this report that we entered the project with more scepticism than belief, though we felt it was a matter which had to be investigated. As we proceeded we became more and more convinced that release of atomic energy on a large scale is possible and that conditions can be chosen which would make it a very powerful weapon of war. We have now reached the conclusion that it will be possible to make an effective uranium bomb which, containing some 25 lb of active material, would be equivalent as regards destructive effect to 1,800 tons of T.N.T. and would also release large quantities of radioactive substances, which would make places near to where the bomb exploded dangerous to human life for a long period.[1]

It went on to recommend that, in spite of the large expenditure which would be necessary (£5,000,000 for a U235 extraction plant alone), the material and moral destructive effect of a bomb would amply reward the effort and cost involved. It forecast that the material for the first bomb could be ready by the end of 1943, and yet again emphasised that whatever lines capable physicists in Britain were following might be precisely the same lines being followed at the same time by German scientists. Its conclusions and recommendations were:

> (i) The committee considers that the scheme for a uranium bomb is practicable and likely to lead to decisive results in the war.

(ii) It recommends that this work be continued on the highest priority and on the increasing scale necessary to obtain the weapon in the shortest possible time.

(iii) That the present collaboration with America should be continued and extended especially in the region of experimental work.[2]

The Use of Uranium as a Source of Power report came to the conclusion that this was a long-term development and was not worth serious consideration as far as the present war went. However, it did recommend that since the results of Halban and Kowarski's work might well have some bearing on the uranium bomb problem, then the two Frenchmen should be allowed to join the workers on the power problem in the U.S.

Make the bomb at top speed; this was the urgent message of the members of the Maud Committee after their 14 months or so of work. What had they seen happening in the world during these months?

For civilians living and working in Britain, this had been one of the most tumultuous periods of all their lives. For the older men the prospect of total war was not new, but that it should have been brought to the Channel doorstep by land and across this doorstep by air was an altogether fresh experience which excited new emotions, provoked new uncertainties of mind and sharpened new attitudes of loyalties. Newspaper reports could hide the nature but not the rate of German progress in France. Tanks were sweeping up through the countryside at a pace believed to be upwards of 40 miles a day. Even accepting the possibility of a breakdown of resistance, this was a speed of military movement which was inconceivable in the war fought 25 years ago in the same country. It was a new kind of war with uncertainties that the Allied force which could be mounted would have the same look of prepared modernity about it as its German opposition. Hitler had marched into the Low Countries, and only two days before the surrender of the Belgian army on 28th May 1940 a congregation of Englishmen had gathered with the Prime Minister in Westminster Abbey for an apprehensive service of Intercession and Prayer. Within a few days civilians from the South Coast ports, harbours and holiday resorts were putting out in inadequate small boats to bring back British and French troops who had been pushed into the sea at Dunkirk. So inappropriate were the means of escape that orders had been issued to rescue only fighting men and leave behind the wounded.

The question throughout the summer was not whether an invasion would come, but how it would be repulsed, and when in August the bodies of German soldiers were washed ashore on the south-west coast of England, rumour had it that an unsuccessful invasion attempt had been made. During September the people of the towns and villages of southern England had watched what the country's leader called one of the decisive battles of the world being fought out. But even accepting the crucial success of the Battle of Britain the danger from the air was far from being over. Professors of chemistry, immigrant physicists, and scientific advisers to the government all took their places in the fire-watching, A.R.P. and A.F.S. rotas along with the foundry workers, the school teachers and the farmers, to watch the attempted destruction of the cities by bombs. In November they helped dig a mass grave for the dead of Birmingham and in December they ran to save what they could of St Paul's Cathedral; in the spring of the following year they were pulling out bodies from homes near the ports and shipbuilding yards rained on by incendiaries and hearing of the single bomb which destroyed the House of Commons. On 12th April, as Winston Churchill was conferring honorary degrees at Bristol University, nearby buildings were still burning from the previous night's air-raid: the Prime Minister noticed that some of the clothes worn by those taking part in the ceremony were still wet and dirty after their rescue work of the previous few hours. One of the foreign visitors to whom he gave a degree on that occasion was Dr J. B. Conant: an American with keen interest in the progress of nuclear fission. During that year over 40,000 civilians were killed and over 50,000 seriously injured by enemy air attacks.

This was the year's backdrop of history which Britain's first-ever group of atomic weapons scientists had lived against. Is it therefore surprising that the leading physicists in the country who had not been spirited away to 'more important' work, once having been convinced that this great weapon was feasible, should commit themselves fully to trying to bring about its birth? Many years after the weeks in which the potential of the atom became clear to them, some of the physicists who were involved with the weapon have suggested that, although they committed themselves without reservation to the bomb, in their heart of hearts they hoped it would never work. How consciously such men held this view at the time and how much of it later crystallised in their minds as a result of subsequent events cannot be known. However, what is certain is that unspoken doubts of this kind were the property of only a small minority. The events in Europe of 1940 and 1941 were a vicious spur to those most deeply concerned with the Maud Committee's recommendations. If 1940 and 1941 were only the first signs of what could be expected from Hitler, then turning uranium into the biggest bomb ever known might well have turned out to be the only way of preventing disaster.

But even realising the possibilities of the projected new weapon as a vast explosive with destructive powers many orders of magnitudes greater than then existing high explosives, did the scientists do their duty as the only men with any vision of the full potentialities by pointing out other perhaps wider moral implications? The answer can only be that these men told what they believed. And what they believed were theories as yet impossible to put to the test: theories which to some were as unlikely to turn into reality as were the rumours of the destructive potential of the super-weapon. Frisch and Peierls had estimated that a fifth of all the energy liberated by the new bomb would be in form of lethal radioactive rays, and that a deadly radioactive cloud would follow the explosion. The Maud Committee had not avoided mentioning these same consequences and that the danger from radioactive contamination would persist for days or perhaps for months. Any scientist who had seen the effects of laboratory accidents involving radioactive substances could have been aware that these possible side effects of the bomb might well cause it to be looked on as a frightful chemical weapon. Then there were the biological effects, though these were not written of at the time. It was known that ionising radiation could affect living organisms. How would these affect human beings? What might be the genetic consequences for the children of the stricken survivors of the first atomic bomb?

In 1940 and 1941 some of these questions were asked and some of these doubts were mooted. But the times – the events of those years – made them subsidiary to the main questions and doubts. Could the atomic bomb be made at all: and if it could would Hitler's scientists make it first? Those scientists most concerned with the bomb had not avoided the moral issue, but the fears, the uncertainties and the fearsome urgency had for the time being driven it into the background. It was inevitable that only the passage of time was necessary to move it back into the foreground where it belonged.

In August of 1941 Lindemann, now Lord Cherwell, sent a memorandum to Winston Churchill telling him how most scientist believed that the new weapon could be ready for use on an enemy within two years. But here Cherwell did not share the view of most scientists. He himself was not prepared to lay odds of '2 to 1 against or even money'. As it turned out he could have made a useful profit at the expense of his fellow physicists. Churchill, however, neglected his statistical adviser's odds and immediately called for action and appointed Sir John Anderson as Minister to take charge of the project. It was fated never to be completed in Britain during the war.

During a cross-examination an attorney once began a question to James B. Conant: 'Dr Conant, as a distinguished scientist and scholar . . .' 'I am not a distinguished scientist,' interrupted Conant, 'but I am willing to be considered a scholar. Thank you.'[1]

Conant was a blunt man, not concerned with niceties for niceties' sake, who saw the vital importance of the role of the scientific administrator in the application of the new technologies to war. 'You are after elephants with a pea-shooter' was one of his derisory expressions. Once he used it to describe his dissatisfaction with the scale of a pilot plutonium production plant he saw being planned as part of the U.S. atomic bomb effort. It was this driving feeling of American scientists and engineers, no matter what their standing, to think big and to know that their country had both the wealth and the technological resources to carry out their plans for huge projects which European countries could only half-heartedly attempt, which made it inevitable that America would emerge as the clear leader in the race for a nuclear weapon. The United States' relative insignificance as a contributor to world science lasted only as long as practical men in Cambridge, Göttingen and Paris were able to keep pace with the progress of atomic physics by their ability to knock together their apparatus at the glass-blowing bench or with the help of a soldering iron. When in 1932 Cockcroft and Walton bombarded lithium with protons and artificially 'split' the atom for the first time, the bill for the apparatus came to £500. The sum caused some mutterings of surprise, and particularly from Rutherford who had been used to seeing his major discoveries coaxed from pink-string and sealing wax constructions costing only a few pounds. But at the same time as Cockcroft and Walton were putting together their apparatus, Ernest Lawrence in the United States was building the first of the cyclotrons which soon reached huge dimensions with costs to match. By 1938 Lawrence and 26 of his staff at the University of California Radiation Laboratory were able to pose for their annual photograph comfortably grouped in the yoke of the 60-inch cyclotron magnet. From here on, if the rapid rate of progress of atomic physics was to be maintained, then the ability to think big was *sine qua non*.

Conant was one of two scientific administrators who were to have a major influence on U.S. nuclear weapons policy. The other, the man to whom Conant acted as deputy, was Vannevar Bush. Just as in Britain as the Second World War approached and the Government began to make unprecedented use of scientists turned administrators, so in the same period did the Government of the United States turn to men with special and specialised qualities. The backgrounds of Bush and Conant, moreover, bear some comparison with those of Lindemann and Tizard. Bush had been a professor of electrical engineering and in 1925 had built the first analog computer. His experience in the administration of scientific research came from his positions as Vice-President of Massachusetts Institute of Technology and President of the Carnegie Institution in Washington. Conant, who had had previous experience of the applications of scientific principles to war as a major in the Chemical Warfare Service in World War I, was President of Harvard University. Bush did not have the almost permanent intimate contact with Roosevelt which Lindemann had with Churchill, and this might well disqualify him from rating as the scientist with more direct power than any other in history, which C. P. Snow awarded to Lindemann. Nevertheless, in the opinion of the Nobel prizewinner, Arthur Compton, one of the atomic scientists most closely concerned with the progress of the atomic bomb, Bush '. . . was responsible for rallying the scientific strength of the United States . . . his great achievement was that of persuading a government, ignorant of how science works and unfamiliar with the strange ways of scientists, to make effective use of the power of their knowledge'.[2] And it was Conant who, again according to Arthur Compton, saw more clearly than others '. . . the magnitude of the operations that must be undertaken if atomic bombs were in fact to be of use in the present war'.[3]

When Einstein sent his letter drawing Roosevelt's attention to the potential of uranium as a weapon, there was in Washington no central governmental office which was capable of receiving information on scientific and technological matter which could be turned to national advantage. For this reason it had been necessary for Roosevelt to create an Advisory Committee which would specifically look into the subject of uranium. There were a number of leading scientists who were unhappy about this situation, Arthur Compton for one, and Vannevar Bush for another. These men felt that in the coming months the United States' Government might have cause to regret that it had never seen it as part of its duties to mobilise its scientific resources. It is true that neither the Government nor the people of America had thought to look on war preparations seriously in the late months of 1939. In retrospect it seems out of character in this nation made rich by technology that the organised utilisation of science should have achieved such low priority at this time. It took until June of 1940 for members of the National Academy of Sciences to make felt their case for some action to have science become a part of the traditional American machine of government. It was in that month that Roosevelt appointed Bush as head of of the new National Defense Research Committee. One of its functions would be to apply science to the needs of war. Compton called it the committee 'that welded American science into a powerful instrument for defending the free world'. Among the many heavy new responsibilities to be an arm of that instrument was uranium research.*

Just as new links between science and government had been forged in Britain, so too had they now been forged in the United States. In Britain, however, where scientists found themselves as members of a nation in danger of being overrun by an enemy, uranium research was directed to the immediate need for a new and powerful weapon. But in America, although much of the significant progress in atomic physics had been made there in the past years, the direction of research had not been so urgently

* The history of the United States Atomic Energy Commission is given in *The New World, 1939–1946* by Richard G. Hewlett and Oscar E. Anderson Jr.

pointed. Now, with the new committee reporting directly to Bush, it was given both direction and purpose.

Even up to the middle of 1941 the target for uranium research had still not been realised. The possibility of building a bomb which could scatter radioactive products over enemy territory was being discussed, as it was in Britain. So too was the possibility of developing a uranium pile which could be used as a source of energy for submarines. Lowest of all in the ratings were the chances of turning uranium into the much-talked-about all-powerful bomb. At this stage there were suggestions from some scientific sources that the whole of the uranium project might well be mothballed for a time. It was in July of 1941 that new facts emerged that made views of this kind change. It was on the 11th that Ernest Lawrence wrote a memorandum drawing attention to the possibility of using plutonium as a weapon. One of the problems in making a uranium bomb was that of separating small quantities of uranium 235 from naturally occurring uranium. What Lawrence was suggesting was the use of the much more abundant U238 to produce plutonium, which might then be used in a fast neutron chain reaction to produce what Lawrence described as a 'super-bomb'.

Although this was not a completely new suggestion it was what was needed at this stage to give the American atomic project impetus towards a potentially crucial weapon. There were other spurs. News had been filtering through from Britain by two-way visits of scientists. It was clear that the progress made by the British was far in advance of what the Americans had thought possible. The news was all the more surprising since American physicists now generally believed that in their ranks they had intellects which had begun to rise above those in Europe, whilst they were certain that in numbers they were vastly superior. Advance news of the Maud Committee report which filtered through was full of optimism that the British believed that a bomb could be made and could be of use in the present war. Bush and Conant had already been informed of most of what they could expect to find in the Maud report by the time Professor G. P. Thomson crossed the Atlantic with his final draft to put them completely in the picture with his committee's predictions.

Events had now conspired to encourage the bomb project into action. American physicists re-examined the problem, made their own estimates of the amounts of material required for a weapon and of its likely devastating effects, and confirmed the British work. In the meantime the administrative channels to the President had been reorganised. A new Office of Scientific Research and Development, headed by Bush, had been created, whilst Conant had stepped into Bush's place and was answerable to him as chairman of the National Defense Research Committee. The O.S.R.D. was now able to apply the results of scientific research directly to national defence and Bush as director of an executive Office could personally ask Roosevelt for whatever support he needed. By November 27th Bush had his own encouraging report ready to transmit to the President. Although it still accused the British estimates of being on the over-optimistic side, its message was clear: a concentrated effort investigating whether fission could be turned into the reality of a bomb should be made immediately. There was not long to wait for action from the top. Within a few days Bush was able to gather members of a new administrative group which had been formed within O.S.R.D., called S–1, and which included some of America's most eminent working physicists. Bush told them that, from a special fund provided by Congress for such a purpose as this, President Roosevelt had supplied several million dollars to put at their disposal. The money seemed a huge sum to the practical physicists accustomed to reluctant Government research grants; only a few of those who knew of it realised that it was only the iceberg tip of the expenditure to come. The date was 6th December 1941. Next day was Pearl Harbour. America was committed to war, and American scientists to the nuclear weapon.

The crucial step in one of the most significant experiments of the 20th century has been recorded in a now famous drawing. The drawing, not at all well executed, but somehow managing to capture the excitement, and even fear, of the moment of truth, shows a group of men standing in the gallery of a racquets court, listening and looking down apprehensively at two isolated figures at work. Among the group of men are just recognisable carica-tures of distinguished American physicists who had agreed to involve themselves in nuclear weapon work. One of the two isolated figures, sitting poised at a bank of electronic instruments, is Enrico Fermi. The date was 2nd December 1942 – just a year after Pearl Harbour – and the place a requisitioned racquets court at Stagg Field on the campus of the University of Chicago in one of the most densely populated areas of the city. A few moments in time after that shown in the drawing, the control rods were withdrawn from a pile containing 6 tons of uranium metal, 58 tons of uranium oxide and 400 tons of graphite blocks stacked neatly on the floor where tennis shoes had once chased racquets balls. It was a disquieting but hushed moment with the main activity coming from Fermi adjusting and reading his recording instruments. The sound which the spectators leaned their ears towards was a quiet quickening clicking of counters registering the rays from the atomic pile. It told them that the experiment was a success and that the first self-sustaining chain reaction had been performed. Before many minutes passed, Fermi judged from his readings that the radiation was mounting to a dangerous level for his unprotected audience and so called for the control rods to be returned and the experiment to be brought to an end. One of the physicists handed Fermi a bottle of chianti so that he could celebrate his success in the wine of his own country. One of the watchers in the gallery, Arthur Compton, was now able to tele-phone James Conant and give a brief sentence saying all that was necessary: 'Jim, you'll be interested to know that the Italian navigator has just landed in the new world.'

And so, at the end of that quiet afternoon, controlled atomic power had become a reality. The amount of power generated in the Stagg Field pile was minute: it could scarcely have made an electric light bulb glimmer, and the amount of plutonium it was capable of producing was insignificant by comparison with the estimates of what was required to make a bomb. The effect of Fermi's pile, however, was that of a great reassuring landmark. Not only was it an epoch-making accomplishment which put the energy of the atom at the disposal of man to use as best he was able: it told the few who knew of it in America that their hopes and their theories were entirely justified up to this point. It was a springboard of confidence on the way to the weapon.

Much had happened since that day almost a year before when dive-bombers at Pearl Harbour had, as it were, shouted an amen to the decision to concentrate American scientists on to the task of the atomic bomb. Already by the day of Pearl Harbour the jobs of investigating the possible methods of separating uranium isotopes had been given out by Bush at a Washington meeting: Harold Urey was to develop a diffusion method, Ernest Lawrence a magnetic method and Eger Murphree a centrifuge method. Arthur Compton was given the job of designing the bomb itself. At a lunch after this meeting Compton talked himself into being given responsibility for plutonium production. It was from eager informal beginnings such as these that the growths of the tentacles of the vast new project sprang. Responsibilities were quickly handed out and funds allotted. Within hours the men who had taken the responsibilities were already persuading the ablest young scientists they could find to join the ranks of the new weapon empire. They were choosing sites and laboratories and they were organising supplies as disparate as uranium, without reliable sources of which the whole project would collapse, or penny glass tubing, a shortage of which at this stage could damage a grand design in a tumbling hurry.

It was now that the great fund of physicists and chemists that the United States had at its disposal began to pay real dividends. No matter whether a practical group was to be given laboratory space at Berkeley or at Chicago, or whether a theoretical group was to be formed on the campuses of Columbia or Princeton, the scientifically well-heeled society could and did provide. This ready availability of men to do the job was a vital factor, but those of available resources and of ample risk capital were equally vital. In Europe the chief protagonists had to cut the coat of their new scientific weapon development according to the cloth of the remaining men and money. In Germany the immense technical commitment of a uranium bomb had not yet begun to be realised and apart from the drain of effort to the day to day running of the war, other projects such as Dornberger and von Braun's rockets needed the stiffening of Reichsmarks and of technologists. Britain had similar problems and radar's development had already had priority in the selection of the best available brains. In spite of what the Maud Committee had achieved it was inevitable that the Americans' efforts should now begin to surge ahead in both pace and range. When, in May 1942, James Conant had to consider how far he dare recommend that the U.S. commit itself to what was still only a dream of a weapon, he had to decide which of the isotope separation processes should be backed. Each one required a pilot plant and each one of these must then be followed by a

8 American and British Intelligence officers dismantling the Haigerloch pile in April 1945

9 Professor Sam Goudsmit on the Alsos mission

production plant. If the race to the bomb were to have real meaning then production and pilot plants must be built simultaneously to save precious months of time. And to go ahead on all available fronts would mean the possible investment of the staggering sum of $500 million. As it turned out this figure was only a quarter of that eventually spent on the atomic bomb, but early in 1942 it represented an expenditure on what many still considered to be, not a hopeful dream, but the germ of a bad idea. Only in America could a man like Conant have seriously considered this vast investment knowing the drain on men, man-hours and materials it would involve.

Conant's eventual recommendations to Bush were on a limited front but they were decisive. In spite of the gravity of the commitment all the preliminary work pointed to the fact that this was the time to begin the production of a bomb; within a few days the recommendations had been approved by Roosevelt.

The scale of the effort was now not only realised but had been given the rubber-stamp approval from the top. To cope with it, it was necessary to bring in some body which had the ability to organise a sprawling empire of men and machines. It was a job for the army: from here on the U.S. Corps of Engineers was to have charge of the plants which would build the atomic bomb. By September of 1942 the emperor, the man who was to stay in command of the newly established 'Manhattan Engineer District' until its successful conclusion, had been decided on. He was Colonel Leslie R. Groves: a soldier not at first overjoyed by this new administrative appointment. He was not a trained scientist but, having taken the job and the appointment to Brigadier General which went with it, he felt that his military and technical background was sufficient to cope with these unfamiliar breeds of physicists and chemists who were now his charges. Although relations between science and the army were often to suffer strain in the next three years, there was no question of the emperor striking up a pose in his new clothes; the most Groves would allow of the situation at this stage was that at least 'I wasn't out in a complete jungle as far as I was concerned'.

Even before the entry of the army the complexity of the enterprise was very great. The broad canvas of the United States was already dotted with centres of study, research, development and industry along with proposed new centres, all intended to contribute to the grand design. By February of 1942 Arthur Compton had begun to develop one of the main laboratories at Chicago to devise a method of producing plutonium. It had been given the code name 'Metallurgical Project' and was to be serviced by many other groups working up and down the country. At the height of its activity it was to employ 5,000 people: 2,000 of them directly in Chicago. The projects still in the planning stage promised even greater complexity. It had been foreseen that it would be necessary to build a special laboratory quite separate from the existing centres where, hopefully, the planned-for uranium 235 or plutonium – or both – could be engineered into the weapon. This ultimate stage would itself demand prodigious efforts of science and technology. And before this stage could be reached, Groves would have to consider where and how he could build the pilot and production plants which were to supply this laboratory.

It was within hours of his appointment that Groves got to the site where the crucial industrial phase of the Manhattan Project would hang in balance. It was to become known as Oak Ridge. The features which Groves was checking over in this rugged 90 square miles of Tennessee Valley which had been recommended to him, were chiefly a reliable and large power supply with which to run his plants, and a good water supply to cool them. There were other factors, such as the remoteness from other centres of population and from prying eyes, as well as at the same time ease of accessibility. This poor, depressed farming area filled his bill. There was power in plenty from the Tennessee Valley Authority, water from the Clinch River, and enough solitude to demand the planning of a whole new industrial city from the grass roots of the farmland. The army moved quickly. It was only a matter of weeks before the farmers and villagers, the parsons and schoolteachers and anybody else who needed to be evicted were moving out and contractors, engineers and scientists were moving in.

In the months to come the situation was to be repeated many times on large and small scale. The landscape in little known parts of America began to be changed with speed and with thoroughness by men with a racing mission. Strange brick and metal constructions began to appear in lonely valleys; jobs suddenly became available in communities where there had been unemployment since the days of the depression; and many of the men who appeared to work in the fenced-off and guarded areas were well-educated strangers who could have been college graduates and perhaps even professors.

The mushroom had begun to grow. Gathering under its shade at each stage of its organic development were more and more of the well-educated men – the chemists, the physicists, the mathematicians and the engineers – dedicated to the great task of producing some of nature's heavier chemical elements with the aim of turning them into a weapon. In these early months they had only glimmerings of ideas of how they would overcome the huge practical difficulties they could see ahead. The unknowns were limitless. But this was part of the intellectual challenge. And this was one of the reasons why they were doing what they were doing. When, on 2nd December 1942, the news spread from Chicago that Enrico Fermi had managed to bring off the first self-sustaining chain reaction there was a sense of elation and booming confidence. The experiment was in no way a practical method of producing plutonium; what it had done was to confirm an elegant physical theory. The physicists could now commit themselves to one more stage of the wider aim.

But commitment was much more than the simple acceptance of an intellectual challenge: it implied an acceptance of the possible consequences of success. And these consequences would be bound to involve the taking of human life on an enormous scale if, as was intended, the bomb could be completed during the present war. Within a few days of Fermi's experiment Vannevar Bush had taken his latest optimistic report to Roosevelt which was to result in an all-out production effort being called for from the project. The date when a bomb could be ready for loading into an aeroplane on its way to a target in enemy territory was being haphazardly estimated as being as early as summer 1944, and as late as autumn 1945. Unless, therefore, there was some unforeseeable collapse in the progress of the war, the chances of the atomic weapon being used were frighteningly high: the more frightening when it is remembered that during 1942 Allied scientists had still every reason to believe that their German counterparts were several months ahead in the race.

First it was as a trickle, then as the months passed and the ultimate vastness of the enterprise was realised, it was as a flood that America's scientific *crème de la crème* was mobilised. Many were old, wise and distinguished; most were young, enterprising and still relatively inexperienced. They flowed into the bomb project both when its beginnings were uncertain and its aims inexplicit, as they did when it had already become one of the most massive technological organisations in history and when its cold aim was clearly defined. What was it that convinced these men that they should do what they were doing? And did they question their own actions and their own motives?

Arthur Compton was almost 50 by the time he fully committed himself to the bomb project. He was a man of deep religious conviction who was strongly influenced by his family's long Presbyterian tradition. His father, besides being a professor of philosophy, was an ordained minister. Compton's personal account of the atomic project, *Atomic Quest,* is freely furnished with extensive quotations from the Bible; in it he writes, 'I see Jesus' spirit at work in the world. As I understand the meaning of the words, this spirit is an aspect of God, now alive in men and women, and through them is shaping our world.' His attitudes to war, however, had already been shaped by previous involvement. As a student at Princeton during the early years of the Great War he had been distressed by the ease with which his colleagues were able to drift into war-like talk when news of disasters in Europe reached the campus. So strongly did he feel that military force served no purpose whatever in settling disputes that he was prepared to argue his pacifist case against his father, whose beliefs had so far so strongly influenced his own.

When at last the United States entered that war his attitude to it was completely reversed. It depended on the great faith he could put in his country's leader:

> ... I accepted at full value President Wilson's statement that this was a war to make the world safe for democracy and devoted myself wholeheartedly to the development of airplane instruments. ... The experience of World War I opened my eyes to the fact that a nation cannot of itself determine to remain at peace.[4]

In general the moral attitudes of the middle-aged physicists were as well-formed by previous events as those of Arthur Compton by the time thoughts of a Second World War came to them. But the atomic project's contribution to this war was not to be hammered out, in the main, by middle-aged men. The average age of the scientific staff at the Los Alamos laboratory, for example, was only 26. In 1939 many of the men who were to make significant contributions to the nuclear weapon's development were at the same state of maturity as was Compton in 1914 as a student at Princeton. In 1939 Robert Wilson was still a Ph.D. student at the University of California, committed only to the study of physics: like the young Compton had once been, he was a pacifist.

Wilson took his doctorate and shortly afterwards moved to Princeton University to carry on his academic work there. He had no thoughts of getting himself involved in war work until in November of 1940 his professor at the University of California, Ernest Lawrence, cabled him asking him to join in some discussions. Wilson travelled to Cambridge, Massachusetts where Lawrence was helping to set up a 'Radiation Laboratory', and there listened to a group of British scientists who gave a briefing on the development of radar in England. This briefing had a deep effect on the young physicist. It was given by a group of men who, in addition to performing the scientific duties for which they had been sent, painted a graphic picture of the effect of the blitz on London and its population. They appealed to American scientists to give their help before it was too late, and insisted that whatever contributions they might make would have important outcomes in the war against Naziism.

Wilson was given only one night to make up his mind whether he would commit himself to radar work in support of the British scientific effort. As a pacifist member of a country which had still not entered the war, this could not be a casual decision for the sensitive and inexperienced young man. That night Wilson did not go to bed. The thoughts which passed through and through his mind were not very different from those of many other scientific pacifist idealists whose services were being solicited to help prepare the new tools of war. He considered the exaggerated but frightening reports of the power of Germany in Europe and weighed the chances of a British victory. He turned over in his thoughts all he had read and heard of what the effects on the world of a German victory might be and whether he would be able to reconcile the consequences with his own ideals for civilisation and morality. Overnight the physicist's idealism disappeared; the next morning he agreed to take part in radar development and had committed himself to the war effort.

But working on radar was not to be how he spent his war. Within a few weeks Wilson returned to Princeton to finish off the loose-ends of teaching and research he had left behind him, and there he was approached by the Chairman of his Department, Henry Smyth, and by Eugene Wigner, one of the three Hungarian physicists who had persuaded Einstein to write his warning letter to Roosevelt. Smyth and Wigner told Wilson of their connections with the new uranium project which they considered to be of greater interest and of more importance for Wilson to join than the radar laboratory. The frank and detailed presentation the two scientists made persuaded the young man yet again to change his tack, if not his overall direction. There was no question at the time, however, that a scientist turning from radar to atomic work should feel more deeply involved in war and in killing because he was turning from defensive to offensive weapon research. On the contrary: for Wilson this was a lesser, rather than a greater, implication because he felt, as a result of being familiar with some of the calculations that had been made, that the chances of making an atomic bomb were small, whereas those of building a nuclear reactor as a useful energy source were high.

Wilson's next year was spent at Princeton on cyclotron work in connection with energy production. As physics the work was of high interest and had unusually far-reaching possible applications. However, the nature and the power of what was most likely to be the first application were not wholly realised by Wilson until the visit of yet another delegation of British scientists. One of the members of this delegation was Professor Franz Simon, one of the Jewish émigrés Lindemann had invited to the Clarendon laboratory and now involved in the British atomic work. To Wilson, who had recently steeped himself in a number of Eric Ambler mystery stories and had been amused by a crop of spy films of which *The Thirty-Nine Steps* was one, Simon was the archetypal Middle-European spy. It was only with difficulty that he was able to overcome the inhibitions preventing him from talking freely to the bald, bespectacled scientist with the suspicious accent. Once having sufficiently conquered his immature reservations, Wilson was swept up in the importance of what the visitor had to say. Simon told Wilson of the measurements recently made in Britain which suggested that a very small amount of uranium 235 was all that might be required to make an atomic bomb.

Wilson questioned the data – some of which later proved to be incorrect – was convinced by Simon, and like many others before and after him was swept into the exciting speculation of how to produce a few grams of the uranium isotope in order to have it produce an explosion. The crucial problem of separating U235 from U238 gnawed at his mind for some time after Simon's visit until late one night whilst walking home alone from his laboratory he suddenly conceived a method of separation using ionised atoms of the isotopes. Wilson is today still able to recall the intense feeling of elation which the process of creativity gave him on that evening and of the heightening of the experience when he believed that he had the answer to the atomic bomb in a nutshell. For several weeks Wilson lived with the understanding that he, the ingenuous kid from Wyoming, had within his grasp the means to end the war. He was still only a few months removed from his pacifist attitudes, yet now he was aware that circumstances, and the wish to conquer a problem which physics had presented to him, had led him to want to make a devastating weapon. As it turned out, Wilson's flirtation with a quick solution was short-lived. Eventually it was realised that it would be necessary to have much larger quantities of U235 than Simon's figures suggested, and as a useful method of uranium separation the method came to nothing. As far as he personally was concerned, however, Wilson now had to acknowledge that with his science he was deeply committed to producing weapons of war.

Richard Feynman was 24 years old and working towards his Ph.D. at Princeton when the 26-year-old Robert Wilson walked into his office in the graduate college where Feynman was living and told him of the remarkable idea of separating uranium isotopes he had just had. The news was unsolicited by Feynman as was the secret information that Wilson was working on the new government project. Wilson invited Feynman to join him on the project as a theoretician and to take along his pencil to a meeting being held on the same day at 3 o'clock. Feynman, an ebullient man with great colour but not always delicacy in his method of expression, firmly assured Wilson that he had made no uncertain mistake to share his secret since he, Feynman, had every intention of completing his thesis before he began to apply his theoretical physics. Wilson left the office with the assurance that his secret was safe, but without Feynman. Feynman returned to his pencil, his paper and his Ph.D., but was able to write for only a few minutes before he found himself pacing the room thinking over some of the implications of the conversation. The worry on his mind was that if Wilson's predictions of the relative ease of production of an atomic weapon were correct, then German scientists, who had in their numbers the discoverers of fission, might well be far ahead in the race to the bomb. The fear of the consequences of such a weapon as this in the hands of a Nazi dictator was what drove Feynman to joining Wilson at his meeting at 3 o'clock that afternoon and to committing himself to the atomic project before the completion of his doctorate.

The fear of a German atomic bomb and the consequences for Western civilisation were not at all unusual subjects for physicists' conversations in the common-rooms of many of America's leading universities during this period. At the University of Chicago for example a frequent lunch-time topic was: should Germans be working on a weapon, what would be their most likely techniques, and how could they be sabotaged? Was it likely that they were using graphite in a pile and if so was there any way of introducing some substance as a poison into the moderator?

Like Richard Feynman many of the freshly graduated physicists of these universities allowed themselves to become involved in the atomic weapons project because they felt this fear of Germany intensely, and the need for action. Still, however, the war was apparently far removed from American shores and only the dramatic turn of events that turned war from being a subject for hypothetical discussion into reality provided the final swing of the pendulum of commitment. Volney C. Wilson, a young physicist working at the University of Chicago, was asked by Arthur Compton to report on whether a nuclear chain reaction could be propagated using ordinary uranium. He was a pacifist of apparently set conviction but who was prepared to work on national defence research in times of peace. He was described by Compton as a 'representative of our finest young American manhood'. He handed in his report saying that he believed that a chain reaction could be produced, that it would be terribly destructive, and for that reason wished to have nothing to do with it. This was before

Pearl Harbour. The effect of the Japanese attack on his pacifism was so great that he completely reversed his previous attitude and eventually asked to be taken back on the atomic project wherever he was most needed.

But not all the men who were eventually to work on the American atomic project could be described as specimens of American manhood. The formative years of a significant number of those who were to play a most telling part in the development of the new weapon were spent several thousands of miles away from the influences of Columbia, Princeton or Berkeley. Edward Teller, the Hungarian who had chauffeured his fellow country-man, Szilard, to the meeting from which had emerged Einstein's letter to President Roosevelt, had not arrived in the United States until 1935. He knew Germany well, had taken his Ph.D. there, and even a year after the drafting of the Einstein–Szilard letter, on moral grounds, had still not decided whether he should commit himself to weapon development. As in the case of Compton, it was a speech from a President which made up his mind for him. On 10th May 1940, the day that the Netherlands was invaded, Teller listened as Roosevelt told a Congress of Pan-American scientists in Washington, 'You and I, in the long run, if it be necessary, will act together to protect and defend by every means at our command our science, our culture, our American freedom and our civilisation.'

Teller was one of very few men who knew what Roosevelt knew of Einstein's letter so that these words had singular significance for him. Within a few months he had taken his citizenship, become part of the American culture of which Roosevelt had spoken, and had joined the atomic effort.

Emilio Segrè, another Jewish émigré physicist who had been dismissed from his job in Italy by the Fascist Government, had no moral waverings which left him undecided as to where his duty lay. A Hitler rampant in Europe removed any necessity to discuss ethics. For him, 'Hitler was an incarnation of the devil, as my Indian friends used to say – so that was all that there was to it.' He was pleased to have himself involved in the Manhattan project.

But for one man who was to have one of the most influential roles in the whole of the Manhattan Project this question of moral duty was never so clear-cut nor so uncompromising as that of

Segrè. He was J. Robert Oppenheimer. He joined the atomic energy programme in 1941 at the age of 37. His decision to do this was to affect critically the nature and purpose of the rest of his life. Twenty-five years later he said of this decision, and of the whole of his attachment to the atomic weapons programme, 'I am not completely free of a sense of guilt.'[5]

Oppenheimer's father had been a rich German-Jewish business-man who had migrated to America at the age of 17 and who had married a Baltimore artist. The son had had a brilliant academic record in America, at Harvard and at Cambridge, and had gone on to take his doctor's degree at Göttingen. In Germany he had carried out some distinguished quantum mechanical work with Max Born and later published highly original ideas on deuteron bombardment of the atomic nucleus. At Berkeley in the late 'thirties he had helped to build up one of America's finest post-graduate schools in theoretical physics.

He was an uncommonly widely read scientist and chose for his friends not only fellow physicists, but scientists from other disciplines, classicists and artists. Of his own admission, however, he was strangely disinterested in economics and politics.

> I was almost wholly divorced from the contemporary scene in this country. I never read a newspaper or a current magazine like *Time* or *Harper's*; I had no radio, no telephone; I learned of the stock-market crack in the fall of 1929 only long after the event; the first time I ever voted was in the presidential election of 1936. To many of my friends, my indifference to contemporary affairs seemed bizarre, and they often chided me with being too much of a highbrow. I was interested in man and his experience; I was deeply interested in my science; but I had no understanding of the relations of man to his society.[6]

The events which forced Oppenheimer's apathy towards wordly affairs to turn into an attitude of fierce personal involve-ment was the fascist treatment of German Jews. Some of his father's family were in danger of suffering from Hitler's worst excesses and he saw that he was now in no position to keep him-self aloof from the practical applications of other men's politics. A more than transient flirtation with communism had ended in disillusion by the time the United States joined the war against fascism. But though Oppenheimer had now pushed his links with the far left to the back of his thoughts there were others who kept them well in mind.

It was on an autumn day in 1942 that Oppenheimer drove with General Groves up a dust track hidden in some New Mexican mountains to give the military head of the Manhattan Project his first view of Los Alamos. This was the site they had come to investigate as a hideaway for the new special weapons laboratory where it was planned that the scattered results of the nation's atomic project would be coordinated to make the bomb which was the end product of millions of dollars and millions of scientific man-hours.

The site was not new to Oppenheimer. He and his brother owned a ranch up in the nearby Sangre de Cristo range of mountains and he knew the area well. When the other sites they had investigated turned out to be unsatisfactory he remembered the wooden ranch that housed the isolated Los Alamos school and drove Groves there to form his own opinion. Groves was taken by what he saw. The whole area was the cone of a vast extinct volcano. Nearby long, tall mesas jutted their sharp profiles from between the canyons of the high hills. On the slopes themselves soft patchwork blankets of dark-green conifer and yellow aspen leaves contrasted with the harsh rock formations. This impressive place seemed to answer the needs of Groves as an inconspicuous home for the new project. It did, however, have its disadvantages. The very isolation which recommended the place carried along with it the problems of inaccessibility which the military road engineers would have to face, and there was a serious lack of a water supply. But these disadvantages could be overcome.

Within days Groves had the authority to take over the land and by the end of the year plans were being drawn up to receive the men who would staff the secret military laboratory. It was amongst this countryside of grandeur that physicists such as Robert Wilson, Richard Feynman, Emilio Segrè, Edward Teller and many others would soon gather to begin in voluntary isolation some of the most intensive work of their lives. Soon after his arrival early in 1943, Segrè stood with his colleague I. I. Rabi on the new laboratory site looking out over the splendid view to the Sangre de Cristo range, and discussed the likely duration of the war, and how long they might be expected to be cooped up in their new home. With considerable pessimism Segrè made the remark that after 10 years perhaps even this view would begin to pall.

Not long after his visit with Groves to Los Alamos Oppenheimer had received a letter signed by the General and by James Conant appointing him as director of the laboratory. Much more than for the others who were to be involved, Los Alamos and its purpose was to shape Oppenheimer's whole life. It was to him that the main task of recruitment fell. He began travelling the country looking not only for scientists with direct experience in atomic energy, but also those who were familiar with radar, electronic engineering, inorganic chemistry and any other skills that might be relevant, and persuading them that here was a responsible and urgent task on which the outcome of the war might eventually depend, and for which he needed their enthusiasm and ability. He was asking them to uproot themselves from their homes, their families and their work and put themselves in scientific purdah under a military command for an unspecified period on a project whose feasibility was still gravely in doubt in some men's minds.

He was excited to do fast and well a job which he deeply believed must of necessity be accomplished. From the outset he insisted that those who were to take part should have a full understanding of what was being done, why it was being done, and what the consequences of the work might be as far as could then be known. In his own words,

> . . . we were a true community of people working toward a common goal. I think that irrespective of what was done with it, irrespective of what was to come of it, it was clear that this was a very major change in the human situation, and the people who were involved in it had a sense that they were playing a part in history. We started out by thinking that it might make the difference between defeat and victory, and ended by thinking that it might make a difference between a world periodically convulsed by increasingly ferocious global wars and a world in which there will be none.[7]

The net which Oppenheimer could cast in order to catch his men was wide, but, strangely, not wide enough to grab quite all of those who wanted to be drawn in. The recruitment for Los Alamos coincided with a period when Anglo-American scientific relations were at a low ebb.

It was the British Maud Committee's report which Vannevar

Bush had used to influence President Roosevelt in his support of the American atomic energy programme back in the summer of 1941, and the interchange of information between scientists of the two countries had gone on apace until the middle of 1942. By this time American investment in the project in men, material and results was far outstripping that of Britain, and the British were now being made fully aware of their security-conscious ally's wish for a restricted exchange of further secrets. There were worries on both sides. The problem of who should own the patent rights in atomic energy when the war was over produced some acrimonious discussion; the British in particular, besides being conscious of their prestige, feared that the whole project might be looked on as a United States invention which would give the Americans the rights of international royalties. Fears for military security finally came to a head late in 1942 when Roosevelt learnt of the Anglo-Russian agreement to exchange new weapons and any which might be discovered in the future. Immediately the President endorsed the policy of limiting the interchange of information with Britain on atomic matters.

And so by the early months of 1943 communications between British and American scientists had come practically to a standstill. Now, for what was probably the first time in history, scientists of nations which were not only totally friendly, but also allies in a great war, ceased to communicate their work to each other. If the phrase 'the brotherhood of science' had ever had any meaning, then this meaning had now disappeared.

The impasse was to continue for several months, and not until August of 1943, when the Quebec Agreement called for 'full and effective' cooperation, was the brotherhood restored and British scientists were again able to contribute their knowledge, and also to draw on the great store which the Americans had in the meantime amassed.

Now for the first time the British were able to learn of the New Mexican weapons laboratory, and Oppenheimer, whose relations with the British had in any case always been good, had a new source of skills to draw on. Among the 19 British scientists who worked at Los Alamos during the war were nuclear physicists, theoretical physicists, electronic experts and explosives experts. And among these, as a result of Britain's own attempts to preserve military secrecy in the early days of the war, was a disproportionate number of émigrés. Of the 19, six had distinctly non-Anglo-Saxon sounding names. Before they had left Britain, all the refugee scientists, with the exception of Joseph Rotblat, had taken British citizenship. Ironically, if either the Americans or the British ever needed justification for their respective policies of wartime security, then both would later be able to find it in the fact that one of these names was that of Klaus Fuchs.

By the time the British arrived Los Alamos was no longer a log-cabin school surrounded by a few clumps of trees. In the early days theoretical physicists had turned architects and practical physicists had turned plumbers so that they could all the sooner have a few buildings in which to start their work. But the original concept of Los Alamos as a small physics laboratory had soon been abandoned; chemists and metallurgists had been called in to look after plutonium purification processes which it had been decided to construct on the site, and a strong engineering group had begun research on the bomb design.

The British found themselves added to a dynamic and still growing community. For most of them it was an unusual and stimulating experience. Jim Tuck had been a physical chemist at Manchester before changing his vocation and his University and moving over to nuclear physics at the Clarendon, Lindemann's laboratory. Lindemann had taken Tuck under his wing and when Churchill had called for Lindemann as his adviser, Lindemann had brought in Tuck as his personal assistant on scientific matters. Tuck was therefore already well aware of many of the major applications of scientific manpower during the current war. Nevertheless, the spirit he found at Los Alamos took him by surprise. Rubbing shoulders with men of the calibre of Hans Bethe, Enrico Fermi and John von Neumann was a highly stimulating experience for a young scientist. He said later of Los Alamos:

> It had all the greatest scientists of the Western World, and something I'd never known before: they didn't care who you were, or what you were. All they cared about was what you could contribute and what you had in the way of ideas. This was new to me. I met scientists I would never have expected to have seen in my whole life before.[8]

Work was hard, the hours men spent at it were long – often up to 18 hours each day – and the isolation and military restrictions of the place could be depressing. But for men in love with their science, the impulse of other men's minds was rewarding.

The spirit of Los Alamos was not without its human focus. The fusion into a purposeful entity of this group of brilliant, self-stimulating and often temperamental men living under curious conditions was strongly dependent on the personality of the man who had been appointed its scientific director: Robert Oppenheimer. The community was one of his creation. Most of the leading scientists who found themselves a part of it had come as a result of Oppenheimer's personal pleading. Even though it was eventually made possible for many of them to have their families with them, the privations of life there could have provided fertile soil for obstacles to grow in the path towards the final goal. Many of the privations were of Oppenheimer's own design: the demands on time and effort and duty, the inevitable inequities in laboratory workers' status, the subjection to military restrictions; but the respect his character was able to command demanded and succeeded in getting the unity of purpose and spirit.

Among the older scientists was the recognition of his qualities as a leader, and among the younger men a confidence which his abilities inspired. In particular his habit of making regular tours of the laboratory to familiarise himself with the progress in each stage of the project, and to place words of encouragement here and there, impressed the junior scientists and prevented any feeling of remoteness between them and the higher administration. On one of these occasions he was tackled by a group of young physicists who were becoming irritated by the fact that some of the craftsmen – the plumbers, carpenters and electricians – were on a much better wage-scale than some of the scientists of considerable technical education and academic background. It was an old and difficult problem, but one which did not in any way over-tax Oppenheimer's powers of persuasion and inspiration. By the time he had finished addressing the physicists' group with his quiet, slow oratory, emphasising the ultimate purpose of the work to which he expected them to sacrifice their unique skill, they were ready not only to accept the *status quo*, but to return to the task with a reborn sense of urgency.

And the urgency seemed very real. Still nobody could speak with any assurance of the sort of progress Germany had made towards an atomic bomb. During the period that the V-weapons were being aimed at London, one young Los Alamos physicist, Philip Morrison, kept a short-wave radio set by his bed tuned in to the B.B.C. Overseas frequency. Every morning as he got up he would flick on the set and play with the tuning dial just to reassure himself that the B.B.C. was still on the air, and that London had not been annihilated by an atomic-tipped weapon.

To most of the men working in the isolation of the New Mexican laboratory, the power of what the end product of their labours would be was now becoming recognisable. That it was practicable to produce a bomb which would have a devastating effect on a city as large as London was fully accepted. But there were wider implications beginning to be acknowledged at this stage. The effects of a weapon on this scale were not only going to be terrible and indiscriminating, they were likely to be of a different nature from any yet experienced in warfare. Besides the power of the blast and the generated heat, what were going to be the consequences of the radiation produced in the nuclear reaction? And what of the political consequences? This thing could apparently cause an Armageddon or alternatively it could create the power with which to dominate the world. What were the responsibilities of a nation which would hold this weapon in its hands? What were the responsibilities of the scientists who were at that very moment in the process of creating it?

As 1945 approached, Los Alamos and the men who worked there were moving towards the centre of the stage to become the focal point of the widely spread great project. But although these men might have been expected to be those most urgently apprehensive about the necessity of some consideration of the political and ethical problems involved prior to the completion of the bomb, this was not the case. They were working round the clock on their contribution, as civilians and scientists, to the war effort. The pressures to succeed were immense. The blind competition with Hitler's atomic scientists was one of the most powerful driving considerations: many of those of European origin had relatives who had suffered and who could still suffer. Many of the Americans too had experienced close personal losses in this war; some of their brothers and sons had died in Europe and the Pacific theatre.

10 Robert Oppenheimer and General Leslie Groves in the Alamogordo desert shortly after the explosion of the first atomic weapon

11 Robert Oppenheimer at his Berkeley home soon after his work
at Los Alamos was complete

12 Alan Nunn May after his release from Wakefield gaol. He is
reading *An Introduction to the Laplace Transformation*

The task in hand, by its very nature, sapped much of the constructive thought of these physicists. Looking back, a number of
scientists admit to having been totally absorbed by the fascination
of the physical and technological problems of the bomb. Some,
the younger ones, look back and confess to having been politically unaware. Others feel that they left overmuch responsibility
and trust in the hands of the senior scientists and administrators
who were closer to the wheels of government. The British had
had instructions not to engage in political activities in this
military establishment in the United States, but this apart, their
outlook did not materially differ from that of their American
colleagues.

There were, however, many informal discussions, one scientist
with another, on what the effects on society of their work might
be, and the arrival at Los Alamos of Niels Bohr, after his escape
from Scandinavia in the autumn of 1943, brought some fresh insight from the world nearer the frontiers of war. It was Bohr, the
much respected, even revered, father-figure of physics, whose
work had laid some of the foundations of the great enterprise
which he now joined. There had been well-based fears for his
safety in occupied Denmark, and now his arrival in the New
Mexican mountains brought a genuine sense of relief to many of
the men gathered on the mesa who had come under his influence
and worked with him in the years before the war. Travelling
under the code-name Nicholas Baker, Bohr was soon dubbed
'Uncle Nick', an epithet still used by his Los Alamos colleagues
long after the war was over. Oppenheimer later said of his arrival:

> He made the enterprise, which often looked so macabre, seem
> hopeful; he spoke with contempt of Hitler who with a few hun-
> dred tanks and planes had hoped to enslave Europe. He said
> nothing like that would ever happen again; and his own high
> hope that the outcome would be good, and that in this the role of
> objectivity, friendliness, cooperation, incarnate in science, would
> play a helpful part: all this was something we wished very much to
> believe.[9]

It was Bohr who laid the seeds of constructive political thought
in the minds of several young scientists, just as in the past he had
laid the seeds of constructive scientific thought, and round this
gentle but incisive man a small circle often centred to discuss the
impact of the weapon on human affairs. Bohr's chief concern was

N   177 for the long-term political implications of the atomic bomb and the necessity of some form of international control. He was already planning to bring his worries for the future of a world suddenly exposed to the nuclear weapon, to the attention of the most influential politicians he could contact: Roosevelt and Churchill. Meanwhile he was content to nurture the political consciences of Los Alamos scientists and to exchange his ideas with theirs in whatever spare moments their midwifery on the weapon allowed. In particular the old scientist spent many hours with Robert Oppenheimer and the two grew close in thought. Most of all, the younger man was impressed by Bohr's concern that, as a result of the production of the atomic bomb, the opportunity should not be lost to alter the nature of the Allies' relations with Russia. Oppenheimer felt that Bohr had put his hand on the central issue and that it was more pressing than what should be done with the bomb once it was ready. Though Bohr's stay at Los Alamos was only a relatively short one, these informal discussions had considerable influence on those involved and helped to shape some lasting attitudes.

There was, however, at least one collective formal expression of concern about the future of nuclear energy held at Los Alamos during 1944. Robert Wilson, during his time at Princeton, had listened to a number of meetings at which the effect of science on society had been discussed. They were what he calls 'The Impact of This on That' type meetings. He therefore decided to call a meeting at his cyclotron's group building to discuss 'The Impact of the Gadget on Civilisation' – 'the gadget' being the commonly accepted code-word for the atomic bomb. About 30 people turned up for the meeting and no record was kept of what was said. Wilson recalls being disappointed at the unimaginative level of discussion. But one resolve did emerge from the talk that went on that day. The scientists had heard that a United Nations organisation was to be set up and that an initial conference was due to be held in the spring of 1945. The meeting discussed its fears that the United Nations might be formed either in ignorance of the imminence of a nuclear weapon, or unconvinced that the bomb could be a reality. If, however, an atomic explosion could be brought off before the spring conference, then, and only then, could there be no doubts about its power and its potential for changing the whole character of war and peace. The tenor of the meeting was that the nuclear weapon should mean an end to war, rather than being one of war's enduring instruments. The result of the meeting was a conviction among those who attended that a significant contribution which could be made towards a world peace would be to bring the bomb into being before the United Nations was brought into being. It broke up with a renewed sense of urgency to get on with the frantic job of building the bomb.

Nonetheless the prevailing atmosphere of Los Alamos was one such that meetings and group actions of this sort did not flourish there, and a most influential factor in the creation of this atmosphere was the laboratory's scientific director. Robert Oppenheimer was himself deeply aware of the responsibilities which the nuclear weapon would throw into the lap of the nation which owned it, and could tell the members of his staff that political considerations were being looked after by others in more influential positions. Oppenheimer's chief preoccupation, and the reason why he was employed at the military establishment, was the building of the bomb. It was his concern to see that his scientists worked only to that one end and in just such cases as these meetings he used his powers of persuasion to emphasise their untimely nature and to concentrate every available effort without more time-wasting deviations.

Hypothetical arguments as to what to do with the bomb once it was built could well look very foolish in a few months' time, should it turn out to be the greatest scientific lemon of history. The cost of the promised technological miracle was by the end of 1944 reaching unforeseen proportions. Sums reaching towards 100,000,000 dollars a month were now being spent on the project. The moneys involved were greater by far than any which academic scientists at least had ever had occasion to think of, let alone experience, in connection with their work in the past. By the beginning of 1945 the individual cogs in the great machine combining industry, universities and the military up and down the United States were nearing the fulfilment of their planned purposes. Oak Ridge now began to supply its laboriously produced enriched uranium 235 to Los Alamos. By a tedious and incongruously slow secret system of road and rail, using at times public transport, the precious element was carried cross-country to the New Mexican laboratory. The Hanford laboratory too had reached its production stage and was sending its first supplies of the other extraordinary metal, plutonium.

Los Alamos by this time had slipped out of its research phase into the hard-core task of production and was able to draw in additional scientific manpower from its widely spread sister laboratories which had passed the peak of their activities. A date in July had been fixed to test the first weapon and the feverish activity to overcome the administrational difficulties, the shortages of supplies and the security restrictions which threatened the deadline being reached, began.

The code-name chosen by Oppenheimer for the test was a strange piece of religious nomenclature; he called it *Trinity*: the result of reading the opening stanza of Donne's sonnet:

> Batter my heart, three person'd God; for, you
> As yet but knocke, breathe, shine, and seeke to mend;
> That I may rise, and stand, o'erthrow mee,' and bend
> Your force, to breake, blowe, burn and make me new.

The metal chosen for the test weapon, in the hope that it could give an impressive symbolic representation of the power of the deity on earth, was plutonium.

The way in which the bomb should be detonated had demanded some of the Los Alamos ordnance experts' most sophisticated thought. In the case of uranium 235 the design of a gun method was under way. In this, sub-critical masses of the metal were to be kept apart to prevent a premature atomic chain reaction taking place, and were then to be suddenly detonated by firing one part into another to give a critical explosive mass. Plutonium presented some tricky problems of pre-detonation since one of its isotopes too readily underwent spontaneous disintegration. The method planned in its case was to surround a mass of the element by a layer of conventional high explosive and blow it into a critical shape. This was known as the implosion method. The date planned for its first use was to be 4th July 1945.

The place that had been chosen was an isolated stretch of desert known as Alamogordo in the Jornada del Muerto Valley in Southern New Mexico, a few hours' southerly drive from Los Alamos. The man who had been given command of the test by Oppenheimer was Kenneth Bainbridge, a physicist who before the war had been teaching at Harvard. Bainbridge's task was to build a desert work laboratory to receive the most expensively produced metal the world had known and to chart its disintegration.

On 7th May he fired on site 100 tons of conventional high explosives to serve as a calibration for the instruments sheltering in bunkers three miles from the point of explosion and designed to record the real thing. But assuming that Trinity was to be a success, where would the real thing be used in a real situation?

On the day following Bainbridge's calibration explosion, Germany surrendered. Victory over Europe was complete and the prime motivation which had initially induced so many scientists to drop their research and to turn to weapon production – the destruction of Naziism – no longer existed.

The removal of the fear which had driven so many men from pacifism or simple political disinterest might in retrospect be looked on as the watershed which could have given them the reason to turn their backs on Los Alamos and its purpose. There were few, however, who felt the wish to do this. One of the few was Joseph Rotblat who, watching the progress of the Allied front in the European theatre of war, had already by December of 1944 come to the conclusion that Hitler had made no progress towards the bomb, that Germany was due for defeat, and that therefore the primary reason for his being in America had disappeared. As soon as he had convinced himself about this he asked for permission to leave Los Alamos and returned to England (where he eventually took citizenship).

It was the émigrés, having more reason to feel compromised by the events in Europe, who more than most reacted to the news of Germany's capitulation. It reached Emilio Segrè in the desert where he was working on the calibration test explosion. The workers at Los Alamos, he felt, had well and truly missed the boat, and the rest of the laboratory's function was to him an anticlimax.

But there was no anti-climax for others: Willy Higinbotham, for example, an electronics expert working on vital instrumentation for the weapon, had already lost two brothers in the fighting in Europe; two more were at that moment fighting in the Pacific. If, to many Americans, Japan had not been the prime enemy in the past, then she was elevated to that role now.

However, there are some who look back and, with honesty, regret that they did not withdraw after V.E. day, recognising that the reason they had used to persuade themselves to become involved in nuclear weapon research had been a fear of Naziism and the spread of Hitler's gangster methods. With the removal of this fear the same men continued to work to overcome the great technological challenge of the bomb with unchanged application and fervour.

Robert Wilson was one of these men. The events of that period, when very young scientists were conscious that they were helping pull on the strings that moved history, have left a deep impression on Wilson. He had heard from discussions with Oppenheimer that a decision about how to deploy the new bomb was in the making. At this news he began a heated argument with Oppenheimer in order to try to convince the man he believed was in a position to influence the decision that the weapon should be demonstrated to Japanese observers as a bloodless means of securing a surrender.

Wilson, having failed to move Oppenheimer, took his idea to other physicists, and one colleague pointed out that, should the demonstration bomb fail to explode, then the painstakingly maintained secret, and all that its surprise could gain, would be lost; such a disastrous result might even serve to stiffen Japanese resistance when the observers returned with their news. Wilson's response was to suggest meaningfully that, 'Perhaps they needn't return'.

It was a remark which Wilson, the one-time convinced pacifist, was horrified to hear himself make. But he now had cause to reassess his old values. In his own words:

> I made the suggestion fully aware of the difference between a first use of a new weapon in which the blood of hundreds of thousands of men, women, and children would be on our hands, and of the dreadful precedent that would thereby be set – and that of illegally retaining, even executing, if you will, a few men. I realised by the dramatic character of the statement how far I had departed from the gentle pacifist of my student days. My regret in retrospect is not for the savagery of the remark, but rather that I did not use more forceful arguments with Oppenheimer in order to influence him to take more seriously my suggestion that some kind of a demonstration be made instead of the actual use. . . .[10]

But the time in which scientists could stop, stand and think was running out; the atomic bomb was near to being born. By 1st

July Alamogordo was prepared, and within hours the plutonium, which was to be its core was ready at Los Alamos to be carried to Trinity.

It was delivered on 12th July, and two days later the gadget sat complete on top of its 100-foot metal tower on the desert plain, ready to perform.

The following day, a Sunday, in the late afternoon, three buses set out from Los Alamos carrying a group of the scientists who had contributed much to the weapon, but who had not been involved in the last-minute preparations at Alamogordo. Richard Feynman just managed to make his bus. He clambered into it from a car bringing him from the airport where he had landed after a hurried flight from New York. He had had a coded telegram telling him the date and time planned for the shot, and had let nothing stop him getting to see the product to which he had devoted the last few years of his life.

By the time the buses arrived at the Trinity site a deep velvet night had spread over the desert. The main activity of the preparations was over, and round the different observation bases men were trying to catch a few hours' sleep; zero hour had been delayed an hour or so after the scheduled 4 a.m. The men who had gathered there represented many phases of the project. There were Vannevar Bush and James Conant, who had been responsible for winning presidential support; Sir James Chadwick, the distinguished discoverer of the neutron, the properties of which were so hopefully to be exploited within a few hours; Richard Feynman, who had not even finished his Ph.D. when he became involved in the affair; Otto Frisch, who had helped give nuclear fission its name; General Leslie Groves, the military head whose bluff methods had often infuriated and often amused the scientists under his wing, but who had produced results; and there was Robert Oppenheimer, the scientific head of Los Alamos, thin-faced and pipe-smoking under a wide-brimmed hat, admittedly worried by his responsibilities and with as yet unrecognised ones to bear.

Rain had come at midnight, but it had cleared after a couple of hours. Shortly before 5 a.m. the weather report with wind directions looked favourable and Oppenheimer and Groves took the decision to go ahead. Kenneth Bainbridge, with a small party under the metal tower, closed some final switches he had kept under lock and key and headed back to the safety of one of the observation posts.

At the base camp the observers lay as instructed, face-down away from the blast. A count-down began over loud speakers, and at 45 seconds before zero a mechanically operated switch was thrown.

As the device was triggered, Donald Hornig put his hand on the one switch which could stop the test up to the moment of actual firing, and had no thought for the spectacle; his concern was to watch the needles of the instrumentation which would warn him of any flaws in the system. It was imperative to preserve the plutonium for a second attempt should the detonation abort.

Richard Feynman was not doing as he had been told. He was a little disappointed to have been put at an observation site on a hill about 20 miles from the explosion. When he was handed a pair of dark glasses by a safety officer and was told that these were for protection against the ultraviolet radiation of the explosion, he reasoned that at that great distance ordinary glass would protect him quite well enough. He therefore stood behind a truck windscreen as he listened over the radio to the count-down approach zero, and so probably became the only man to watch the first atomic bomb explosion without the protection of dark filters.

For all its 20 miles' distance, it took him by surprise; it was as though a flash-bulb had been set off in front of his face. In fact the light intensity of that flash could conceivably have been visible from another planet. The shock made Feynman look down to the darkness at his feet, and there he saw a beautiful purple sphere expanding – the after-image of the explosion on his retina. Looking up again, he watched a bright white light turning to yellow clouds; this he knew could not be seen by those around him because of their dark glasses. Gradually the yellow emissions grew darker until, in the distance where the purple sphere must seconds ago have been, an orange ball grew, then disappeared into grey smoke. A glowing halo surrounding the clouds made Feynman once again believe that a second spurious image was affecting his retina, but turning his eyes for a moment told him that this was not so. The clouds were beginning to cause the surrounding air to emit a radioactive glow.

All this, at a distance of 20 miles, had taken place in absolute silence. When the sound arrived, with a crack and a thundering roll which echoed from the mountains, the tension of the watchers was released. There was a gasp of elation. People jumped for joy.

Robert Oppenheimer looked at the rising cloud of the spectacle he had christened from the stanza of John Donne and, as he was later many, many times to recall, found himself again providing other images from his readings; now the Sanskrit of the Bhagavad Gita which he had studied during his leisure hours as a Professor at Berkeley:

I am become Death, the destroyer of worlds.

Oppenheimer was not the only man to look solemnly over the Alamogordo desert and wonder whether man had reached that point where he could usurp the creator's prerogative. The bus-trip which took the scientists back to Los Alamos was a very different journey from the optimistic, chatty journey down to the desert. The bottle of Scotch Willy Higinbotham had carried out to celebrate the success of Trinity was drunk in gloomy silence, 'each of us groping to understand what we had witnessed'. Jim Tuck sat in the group of muted and more thoughtful men, profoundly disturbed by his experience of watching the easy release of such great power as the response to a man-made signal. Tuck began to repeat to himself the question, 'What have we done?'

Long after his work at Los Alamos was complete, Robert Oppenheimer said, 'When you see something that is technically sweet, you go ahead and do it and you argue about what to do about it only after you have had your technical success. That is the way it was with the atomic bomb.'

And that is the way it was with the men who worked in his laboratory on the bomb. So intent were they on extracting the sweetness of perfection from their creation, so devoted to the race against unknowable competition in Germany, and so prepared to let no consideration, whatever it might be, hinder their progress, that arguments about what to do with the bomb once it was made, were not, during its making, at the forefront of their minds. It was from outside Los Alamos that the most anxious glances were being cast to see what was going to be done with the sweetness of success.

The race to the bomb was a relay-race with Los Alamos scientists carrying the baton in the crucial final home-stretch. Many men from many other laboratories had held it in the early laps, but now, having passed it on, they were able to stand to one side, take breath, and watch. It was in these watchers, men who were scientifically very well acquainted with the development of an atomic bomb and its potential power, but on whom the pressure of the moment was no longer fixed, that there developed some of the most agitated concern about the consequences for the world once the race was won.

Niels Bohr had held the baton early. His work with John Wheeler in 1939 had been published two days before the outbreak of war in Europe and still stood as a classic exposition of the mechanism of nuclear fission. Bohr had deduced that naturally occurring uranium, with its high proportion of U238, was much less likely to produce a chain reaction than pure U235. This postulate had been a highly significant step towards the bomb. But Bohr's stature in the world of physics was immense long before the possibility of nuclear fission was foreseen. Earlier in the century he had seen how it might be possible to combine Planck's quantum theory with the existing knowledge of the internal structure of the atom and so explain how energy was emitted and absorbed by various elements. For the first time Bohr was able to provide a satisfactory explanation relating the lines of a spectrum and atomic structure. This work was a great milestone in physics, and indeed for all other branches of natural science, and for it Bohr was awarded the Nobel Prize of 1922.

This, and other great leaps in intellectual thought, established Bohr as one of the accepted leaders of the world scientific community, and the laboratory he had built up at Copenhagen became as obvious a stopping-off point during the careers of physicists as were Cambridge or Göttingen. He was a natural teacher who easily slipped into the role of father-friend with his pupils, and whilst he could draw from them an admiration of the unity of his personality and of the superiority of his intellectual abilities, he was at the same time able to enthuse them with, as one of his pupils described it, a 'fellow-conspiratorial' attitude to work and life.

If it can be said of other leading scientific figures of the early 20th century that they were politically disinterested and reluctantly let events draw them into social action, the same cannot be said of Bohr. In the 'twenties he had done what he could to re-establish Germany in international scientific affairs and in the 'thirties he had used his stature as a world scientific figure to make a direct appeal to Stalin on behalf of a Russian physicist in trouble with the régime. Frequently, he talked politics with his pupils and it seems natural that Heisenberg should have turned to Bohr in 1941 in the hope of discussing the nuclear weapon.

As we have seen, the result of the German scientist's visit to occupied Copenhagen was to trouble Bohr deeply, and only served to reinforce his decision to give away nothing that might be of use to the Germans, either scientifically or politically.

Relations with the Danes and occupying forces worsened considerably during 1943, though Bohr managed to keep his Institute functioning. Then, on 28th August, an ultimatum demanding the death penalty for saboteurs caused, as it had to, the downfall of the Danish government, the proclamation by the Germans of martial law and the beginning of a period of persecution and fear in Denmark. Bohr's uncompromising attitude to Naziism had not gone unnoticed and news filtered through to him that he and one of his sons were due for deportation at the same time as the inevitable Gestapo purge on Danish Jews was to be put into operation. A month later ships were lying at anchor in Copenhagen harbour waiting to take on their first consignment of human cargo for Germany.

Bohr was not added to the shipment. On the same evening that he heard his arrest was due, he and his wife were climbing into a small motorboat on an empty beach lit by an unwelcome moon. Within hours he was in Malmö, and the safety of Sweden.

Bohr had already had one message from Britain inviting him to work there. It had come from Sir James Chadwick through British Intelligence, as a piece of microfilm less than a millimetre square, hidden in a bunch of keys. In it Chadwick had said, 'There is no scientist in the world who would be more acceptable both to our university people and to the general public.' Whilst Chadwick's message was perhaps intended to flatter as well as to reassure, the wording shows a little of the insularity of the scientist. University people would undoubtedly be delighted to have Bohr amongst them again, but it is a sad fact that a statistically insignificant proportion of the general public would even have heard of this, one of the greatest and most influential physicists of the century.

Lindemann, continuing his policy of mounting a collection in Britain of distinguished scientists from the rest of Europe repeated Chadwick's offer to Bohr in a cable to Stockholm. By 6th October Bohr was beginning the next stage of his Odyssey. The manner in which he did it was to become, like the story of the origin of the title of the M.A.U.D. Committee – so unwittingly named after the Bohr family governess – a part of anecdotal scientific history. The unarmed Mosquito bomber which carried the diplomatic pouch between Britain and Sweden was ordered to return carrying Bohr in its only passenger space: the empty bomb bay. Bohr, fitted with a parachute and a couple of flares, in case the worst came to the worst and he had to be deposited in the North Sea, was also given a flying helmet and earphones, and instructions to wait for the pilot's signal to fit an oxygen mask when the plane reached its cruising height. Remissly the pilot had not reckoned with the size of the intellectual giant's head. Bohr was unable to fit his helmet properly over his great dome, failed to hear the orders to take in oxygen, and fainted. Getting no reply on his inter-com, the worried pilot reduced his flying height for the rest of the journey, and by the time Bohr had regained freedom and Scotland, he had also regained consciousness.

In London Bohr learnt to his surprise of progress in the atomic bomb project and agreed to join the British team; in November, with his son Aage, who had also made his escape from Copenhagen, he sailed for America to see at Los Alamos and Oak Ridge the reality of the hardware of the Manhattan Project, the scale of which he had not believed possible until the past few weeks. And in America what he saw increased in him the sense of political urgency which he had now developed since hearing of the bomb's progress. He felt that the strictly maintained secrecy at that time surrounding the project would inevitably breed suspicion of the ultimate motive. Unless some form of communication were set up at this stage, and not later, then the chances of a nuclear arms race developing, as soon as the Eastern bloc's physicists had had time to catch up with their research, were frightening.

In April 1944, Bohr set out for London, but in the few weeks that he had been in the United States, he had managed to achieve more than to formulate some of his own ideas, try them out on Los Alamos scientists and graft their ideas on to his own. He had managed to make high-level political contact. Through acquaintances in diplomatic circles, both American and British, he had learnt that Roosevelt shared his concern for the wider implications of nuclear energy and recognised, not only the grave dangers inherent in the existing situation, but also saw in it some singular possibilities. And so when Bohr arrived back in Britain he was carrying a message that had been passed on to him, that the President would be interested to hear Churchill's views on the wider implications of the atomic bomb.

Bohr arrived in London full of optimism. He had not talked

189

with Roosevelt, but this chance of acting as a catalyst to the two Heads of State would be a unique opportunity for him as a scientist to influence world affairs because of his intimate understanding of the possible consequences of the latest piece of scientific invention: an understanding which perhaps politicians might never otherwise fully appreciate until too late. Bohr sat and waited for the call to Churchill.

Churchill's views of scientists were as subjective as they were of other men. Among the older breed he knew Lindemann well, liked him from the moment he first met him at a house party in 1921 and trusted him sufficiently to have him as one of his personal advisers. Some of the younger ones too had struck a responsive spot. When, in 1940, R. V. Jones was called in to a special meeting in the Cabinet Room to explain the principle of the suspected German Knickebein radionavigational beams Churchill was immediately impressed by what he saw of the young man. In *The Second World War* Churchill wrote: 'For twenty minutes or more he spoke in quiet tones, unrolling his chain of circumstantial evidence, the like of which for its convincing fascination was never surpassed by tales of Sherlock Holmes or Monsieur Lecoq.'[1]

But for scientists in general the Prime Minister held no special brief, and on the subject of the disclosure of atomic secrets to Russia, he later said in Parliament: 'Whatever may be decided on these matters should surely be decided by Parliaments and responsible governments, and not by scientists, however eminent and however ardent they may be.'[2]

If such a firm attitude existed before Bohr tried to see Churchill then Bohr had no foreknowledge of it. The meeting, when it eventually came, was a disaster.

Arranging an interview with a Prime Minister carrying the cares of a nation at war was not easy. Bohr had to wait around anxiously until 16th May before the combined persuasion of Field Marshal Smuts, Lindemann, and Sir Henry Dale, the President of the Royal Society, had managed to get Churchill to agree to see the scientist. The unhappy course of the interview has been recorded by Margaret Gowing, the historian of the United Kingdom Atomic Energy Authority, in a series of interviews with Bohr shortly before his death, and by his son, Aage, who travelled with his father on the shuttlecock journeys between Britain and America.

From the start at Downing Street things went badly. To Bohr's great distress the main issue, the reason why he was in the Prime Minister's presence at all, was lost in a backwater as a result of an argument on Anglo-American collaboration which developed between Churchill and Lindemann, who had joined the meeting. Worse, Churchill, in spite of the qualities other men quickly saw in Bohr, was clearly not impressed by the character of this, one of the scientific establishment's most distinguished representatives. It is possible that Bohr's manner of speech – a rambling, throaty intimate whisper which forced his audience to crane forward to hear – alienated the listener on this occasion, who was in any case so quick to be influenced by subjective qualities. Churchill was not the only man to find Bohr difficult to follow. There were many others. After the war Emilio Segrè, the physicist who had known Bohr at Los Alamos, visited Copenhagen and listened to a description of this very interview with Churchill, but found Bohr's conversation too difficult even for him, a sympathetic scientist, to understand the nuances of what went wrong. It might well be that this alienation of one great man by another was due to the fact, as Cyril Connolly not unreasonably deduced, that Bohr was a bore.

Whatever the cause, Bohr that day was never able to express his views and found himself leaving the room before having reached the point of beginning to discuss the need for international control of nuclear weapons. Bohr, as a last gesture of hope, turned to ask if Churchill would object if he wrote a memorandum to him. Churchill's acid reply was, 'It will be an honour for me to receive a letter from you – but not about politics.' Clearly, this scientist had produced no Sherlock Holmes or Monsieur Lecoq fascination. So ended, in memorable lack of communication, the meeting of highest level science and politics.

A much-saddened man, Bohr could at least begin to plan for his return trip to the United States, which had had to be delayed as the days spent waiting for the meeting with Churchill had dragged by. Before the end of the next month Bohr was in Washington and there received the news to restore his spirits, that Roosevelt would be willing to see him. Roosevelt asked for a preliminary memorandum to be prepared and this Bohr dictated with the customary method of careful revision and counter-revision for which he was well known among his co-workers. His secretary on this occasion was his son, Aage, who sat and typed whilst his father darned the family socks.

In his memorandum Bohr made a plea for action to be taken. The wartime alliance with Russia still existed, and the time to act was soon. Russian scientists would, with the present war off their hands, be easily capable of competing in the production of atomic weapons. Even an approach on a non-technical basis would remove inevitable suspicion and pave a smooth path for a future rapport. Perhaps bearing in mind his previous interview with a statesman, Bohr wrote,

> Of course, the responsible statesman alone can have the insight in the actual political possibilities. It would, however, seem most fortunate that the expectations for a future harmonious international cooperation which have found unanimous expression from all sides within the united nations, so remarkably correspond to the unique opportunities which, unknown to the public, have been created by the advancement of science.[3]

Bohr was also concerned to put to Roosevelt the view that the communication without barriers which scientists had had in the past might again be used to provide the basis for new links in international relations, perhaps on a personal level.

In fact Bohr's personal contacts with Russia were surprisingly fresh. During his last visit to England he had received a letter via Sweden from a Russian physicist he had once known well, Peter Kapitza, the distinguished pupil of Rutherford at Cambridge. The motive for Kapitza's letter was ironically similar to that which British scientists had been slightly quicker to think of: that of inviting Bohr and his family to Russia to live and work. Bohr, after consultation with the British secret service, replied without committing himself beyond the usual amities and felicitations with which scientists customarily greet each other in letter form.

To Bohr's joy the meeting with Roosevelt at the White House took place late in August; its atmosphere could not have been in greater contrast to that which had taken place at Downing Street three months earlier. For over an hour Bohr had the ear of the President and was able to develop the theme of the memorandum. He left full of optimisim that Roosevelt was with him in spirit, and the promise that when Roosevelt met Churchill in the near future these problems would be on their agenda for discussion.

Once more Bohr was to meet bitter, even humiliating disappointment. The circumstances under which Roosevelt and Churchill met in September 1944 at Roosevelt's home, Hyde Park, to discuss the atomic bomb will perhaps never be fully known. Even the outcome itself, an *aide-mémoire,* remained a mystery for some months to those in the United States most concerned with the applications of nuclear energy. Though a photostat copy of Churchill's original document was eventually sent to America, Roosevelt's copy was not discovered until many years later in a naval file where it had apparently been placed because its concern with 'Tube Alloys', the code-name for the British atomic project, was thought to refer to some piece of ship's hardware.

It was to Niels Bohr that part of the *aide-mémoire* referred. Besides stating that the world should not yet be told about the possibility of 'a "bomb" ' and that full collaboration between the United States and the British Government in developing atomic energy should continue after the defeat of Japan, it recommended that some enquiries should be made about the nature of 'the activities of Professor Bohr' and that steps should be taken to prevent him from leaking information to Russia. This was what Robert Oppenheimer later called the tragedy of the perversion of Bohr's counsel.

Clearly, Churchill had reacted against Bohr's attempted advice even more violently than Bohr had feared and had persuaded Roosevelt at their meeting against any immediate steps towards international control. Staggeringly, Bohr, as a result of Churchill having learnt of his correspondence with Kapitza, had been noted as a possible security risk – or worse.

Scientists and non-scientists soon muscled-in to convince both Churchill and Roosevelt of Bohr's integrity and reliability, but the harm was done. In vain Bohr crossed and re-crossed the Atlantic in an attempt to have his views heard on the momentous consequences for the world of the now accelerating atomic project. He still did not understand what it was that had made Roosevelt at Hyde Park make such a violent retreat on the views he had held – and so cordially – at the White House a few weeks before. The parting words which Roosevelt had made to Bohr at this White House meeting were only marginally different in content from those which Churchill had shot at him at Downing

Street, though they were worlds apart in design. They were that Bohr should feel free to write at any time. And this, now that all else had failed, is what Bohr decided to do. He had completed his new memorandum by early April of 1945. He was looking for the best channel by which to have it delivered to the presidential desk when he heard of Roosevelt's death.

Above all else, Bohr was preoccupied by the idea that a revolution was about to take place in the application of science to warfare and that, at that moment in time, an unrepeatable situation existed which could be used to prevent the development of a possibly disastrous tension between the Eastern and the Western powers, not only with respect to international control of atomic energy, but also with respect to many other issues. His vision was large. He had accepted what he believed to be his duty as a scientist possessing special knowledge of a weapon of war, which he in some part had helped create. He had worked alone, and he had failed.

But at the same time as Bohr was making his worried efforts, other scientists whose thoughts, like Bohr's, were not completely dominated by the immediate technical problems of production of the bomb, were also beginning to be troubled by how its use might affect mankind.

As early as the closing months of 1943 the workers at the Chicago Metallurgical Laboratory, where plutonium research was being carried out, knew that their contribution to the bomb project as a whole would shortly pass its peak. Rumours that there would soon begin a sharp reduction in scientific staff, and even that the laboratory might be disbanded, came nearer to distasteful fact a few months later when Arthur Compton, the director of the Metallurgical Project, was asked by General Groves to consider a cut-back which might involve as many as three-quarters of his staff. This was not news to gladden the hearts of young physicists who not only had intently absorbed themselves in an equally young branch of their science, but who were now beginning to see over the horizon to some of the possible applications of nuclear research. So optimistic, and perhaps over-optimistic, was this outlook that one worker felt obliged to dampen the enthusiasm of his colleagues by the reminder that atomic power, in spite of their fine-sounding words, had still not arrived and advised them not to 'talk like magazine ads for postwar plastics'.

But still, the brave new world promised brave new things, and to be ushered out of it after having worked on the exciting phases of its birth seemed a waste of boundless technological opportunities. Many felt deeply the need for the unbroken continuity of their research. There were other problems of a less momentous, but nevertheless unsettling nature. Just as at Los Alamos the young scientists felt dissatisfied with their pay-packets, which they could compare with those of plumbers and carpenters, so at Chicago the scientists and engineers felt badly done-by in comparison with men contracted in from U.S. industry to work at the same level. But whereas the Los Alamos workers had sublimated their discontent in the helter-skelter job of building the bomb, the Chicago workers were facing the first suggestions of redundancy; they had time on their hands: for some, time to be unhappy with their lot and, more important, for others, time to worry about the future for the world which would soon be introduced to the technological creation to which they had contributed.

The last months of 1944 were also bringing a new situation to bear on the thoughts of scientists who were beginning to be nervous about the opening of their Pandora's box. It was clear that, although the final crushing of Germany might take as much as five, six or even more months, the ultimate result was certain, barring Hitler's deliverance by some monstrous secret weapon which had evaded detection by every device of combined Allied scientific intelligence. And this remote possibility was becoming daily more remote as the Allied advance swept up scientific installations which might have concealed unsuspected research. But the cupboards were bare, so that, even if Germany managed to hold out until after the completion of the first bomb, there would be no justifiable reason for putting it to use in that theatre of war. Europe could sit safe; but could Asia? Scientists, like the rest, had to make their own guesses as to how long war in the East might last. If they were to base these guesses on newspaper dispatches and reports from visitors to the war zones, and if only existing weapons were to be brought into play, the future held enough uncertainty to cause very grave concern. Even as late as 15th July 1945, the *New York Times* was publishing its famous headline, 'Newsmen at Guam Guess June, 1946, for War's End'. But even with a pessimistic prediction such as this one, there was no question of the eventual outcome of the war. The question was one of time.

Japan was sitting well in line for the bomb. If the weapon was ready in time to make a significant difference to the Japanese war (and as success in each production stage of the Manhattan project made this now seem a certainty), should it then be used? If it came to that, should it ever be used at any time in any war? Could international control of atomic weapons be introduced, and could such control hope to prevent an arms race?

There were other centres of atomic research where a few men had started to ask this sort of question, but only in Chicago, hundreds of miles away from the focal point of the new weapon's construction, were the critical conditions realised in which considerable numbers of scientists felt the ethical dilemma sufficiently deeply to take positive action. Some worked alone to pose these questions in places where they might be heard, and some worked in groups. Now, for the first significant time in history, scientist members of a nation at war were collectively beginning to question the right of their nation to use a weapon they had created, and to take steps to influence the policy which would decide its use. The likely enormity of the wake of this instrument of war to which they had dedicated their science was what at last drove men of consequence in the working scientific community to lift up their hands and say, 'this far and no further until you've heard what we have to say'.

It was in the last months of 1944 that the first signs of the foment of social conscience began to work under the partial vacuum that had been created, and there came bubbling to the surface a number of ideas with common themes which were pleas for action. Arthur Compton had provoked discussions of the future of atomic energy with many of his staff and had appointed a committee under the chairmanship of Dr Zay Jeffries to make a report for General Groves on 'postwar work on nucleonics'. 'Nucleon' was a word invented by Jeffries to describe collectively the electron and the proton, which are the particles of the atomic nucleus. The report of his committee of six senior scientists pointed out, among other things, the necessity of international control of atomic armaments. At the same time other scientists were persuading Compton to forward to Washington a memorandum calling for the publication of some information on nucleonics which might at least reassure the United States' allies and which might allay any suspicion of her ultimate motives as well as guarantee her future safety.

More than 10 years after the events in these months of ferment and uncertainty at the Metallurgical Laboratory, Arthur Compton published his own personal narrative of how he came to subscribe to the view that there was no alternative other than to use the bomb. Compton had on several occasions listened understandingly to the expressions of the troubled conscience of the physicist, Volney Wilson. It was the same young scientist who at this crucial time played Jiminy Cricket to the conscience of the Nobel prize-winner, Compton.

Wilson, who in 1940 had asked Compton to remove him from the atomic project because he believed he saw terrible destruction ahead, and who after Pearl Harbour had changed his mind and asked to be taken back, now again went to Compton. Wilson's worries, according to Compton, were those of Christian compassion. This time it was too late for a volte-face.

The extent of destruction which the blast and heat of an atomic bomb, if successful, could cause, had long been clear to every scientist working on the project; what was still not clear was the effects of radiation. The severity of skin burns from exposure to radioactive rays was familiar and frightening, but what of the biological effects? Animal experiments had already shown how plutonium emissions were capable of destroying the source of red-blood cells and causing leukaemia; they had also shown that heavy doses of radiation could have some disturbing genetic effects which included the mutation of the genes. The victims of the first bomb would provide the first experimental evidence of the results of massive exposure on humans. Every physicist working on the fission bomb was aware of these potential consequences, though few others outside the scientific world had any notion of them. General Groves, for example, as military head of the project, did not become aware of the post-explosion effects of his embryo weapon until January of 1945.

To Volney Wilson, as to his friend and mentor Compton, it was now clear that the holocaust and the possibility of human suffering he had foreseen and worried about four years ago would soon become terrible fact. Could not some alternative be found to the bomb they were both helping to build, was the question he asked.

Compton, a distinguished and influential figure in the Manhattan project who was nearer to the men who ultimately would decide whether or not to use the weapon, did not record how he answered; he had yet to decide his own course of action.

Wilson was only one of many young scientists at the Metallurgical Laboratory who were agitated. But there were others at Chicago who favoured a more direct path to the top. Among these were older men: some with more experience of the hard ways of the world. Typically, Leo Szilard decided on a personal frontal assault, and lined up his sights on the most distant, but most important target, Roosevelt.

His weapon was to be the same one he had used over five years ago with such success: Einstein. Was the weapon outmoded, however?

Einstein was quite unaware of any progress towards an atomic bomb. During the war years he had on occasions continued to write and speak his views on peace and supranationalism, but he was becoming more withdrawn. To a friend he wrote, 'I live here in great isolation and keep busy with my work. Although I often am on the right track about things in general, I completely lack the ability to speak convincingly and influence others effectively.' But Szilard persuaded him yet again to try to influence Roosevelt. Again the result of the visit was a letter addressed to the President of the United States: this time an attempt to bring Szilard into direct contact with Roosevelt. The last paragraph of the letter, dated 25th March 1945, read:

> . . . The terms of secrecy under which Dr Szilard is working at present do not permit him to give me information about his work; however, I understand that he now is greatly concerned about the lack of adequate contact between scientists who are doing this work and those members of your Cabinet who are responsible for formulating policy. In the circumstances, I consider it my duty to give Dr Szilard this introduction and I wish to express the hope that you will be able to give his presentation of the case your personal attention.
>
> Very truly yours,
> A. Einstein.

It was a vain hope. The letter was found in Roosevelt's office after his death.

The death of Roosevelt, just as it had with Niels Bohr's attempted personal approach, had, for the moment at least, wrecked Szilard's highest hopes of getting the fears of scientists injected into the thoughts circulating in the highest political strata. Undoubtedly, Bohr and Szilard were quite unaware of how nearly their separate paths of action came to meeting on the Presidential desk-top. But within the next few months this was to be by no means the only example of the right hand of scientific conscience not knowing what the left hand was doing. The apparent confusion was to increase. This confusion, however, was a sign that a political movement within science, in little pockets of unconnected activity here and there, was floundering into existence.

The revolutionary nature of the bomb itself, and the new situation which the eased physical involvement had forced on some scientists, had at last made it overwhelmingly necessary for them to have a say in the weapon's use. As its creators, as scientists with an exclusive knowledge of its powers, they felt that they, like politicians or generals, should influence the decision of how to apply it to mankind, and, as a result, they were accepting responsibility for the consequences of their work.

The memorandum Bohr had prepared for Roosevelt, in spite of his humiliating experiences, had not in fact been a wasted effort. He had taken it to the scientist-administrator Vannevar Bush who himself was greatly concerned about the consequences of atomic power in a post-war world. Bush was in a position from which he could closely influence U.S. policy, and was already in touch with the Secretary of War, Stimson.

On 25th April 1945, Stimson paid a call on the White House and the new President who had stepped into Roosevelt's shoes. So closely had the atomic secret been kept that Harry Truman had to be briefed by Stimson on its expectations virtually from scratch. Stimson described it to him as 'the most terrible weapon ever known in human history'.

The restless scientists at Chicago, however, had as yet no way of knowing that the worries they now had at the forefront of their minds had in some form reached the Presidential office. They did not know for example that Stimson had described the consequences of their weapon in strong phrases, and as being such that 'modern civilisation might be completely destroyed'.

What Stimson called 'the proper use of this weapon' was one of the major points he had brought to Truman's attention.[4]

Nor did they yet know that Stimson had also suggested the establishment of a committee of specialists to make recommendations to the President on the questions which the ownership of the bomb would raise. Within a few days of this April White House meeting, the group, which was soon to bear some immense responsibility, was formed. It was called the Interim Committee. Its members were Stimson as chairman; his assistant, George Harrison; James Byrnes as personal representative of the President; Ralph Bard, Under Secretary of State, and, in a numerical minority, three scientist-administrators much respected by the scientific community: Vannevar Bush, James Conant and Karl Compton, President of Massachusetts Institute of Technology (the brother of Arthur Compton). This committee, however, lacked members who directly represented the working scientist and who could express some of the strongly held views of the men who believed themselves to be most intimately aware of the potential of nuclear energy. To correct the imbalance, and to give their advice in general, four physicists were appointed to assist the Interim Committee. They were the three Nobel prizewinners, Arthur Compton, Enrico Fermi and Ernest Lawrence, along with their fellow physicist, without a laureateship, Robert Oppenheimer. They were men who had a wide range of contacts with the inhabitants of the laboratories of the great sprawling atomic enterprise; Compton could in particular represent the views of the discontented at Chicago; Oppenheimer could be expected to know the pulse-rate of the fervent and overworked at Los Alamos, and Fermi, like Lawrence, having the wisdom born of continuous connection with the historical development of the new technology, could in addition speak as a representative of the highly vocal group of émigré scientists.

The first meeting of the Interim Committee with its Scientific Panel of advisers took place in the Pentagon on 31st May 1945. Also attending were General Marshall, General Groves, and two of Stimson's assistants. A day and a half later the Interim Committee had unanimously adopted the recommendations which Henry Stimson later recorded as:

1. The bomb should be used against Japan as soon as possible.
2. It should be used on a dual target – that is, a military installation or war plant surrounded by or adjacent to houses and other buildings most susceptible to damage, and
3. It should be used without prior warning (of the nature of the weapon).[5]

Stimson also recorded that one member of the committee, not a scientist but the Navy Under Secretary, later changed his view and dissented from the third recommendation.

How many of the opinions of the laboratory scientists had been introduced into the discussions leading to these stern historic resolutions? Stimson had opened the meeting by encouraging all of those present to give their views freely on either scientific or political aspects of the weapon. Surprisingly, writing 10 years after the event, Arthur Compton recalls that throughout the morning's discussion of 31st May 'it seemed to be a foregone conclusion that the bomb would be used. It was regarding only the details of strategy and tactics that differing views were expressed.' Oppenheimer, however, did not recollect the attitude of the group which had gathered that day with nearly such clarity as Compton. Twenty years after the event he said to the author,

> The parts of the meeting that I remember, I believe that they were the most protracted, the deepest, the most intense. The ones in which the Secretary of War, Colonel Stimson, and General Marshall participated, had to do with the post-war future, with the prospects – technical prospects – which we were not all knowing about, but of which we knew something, and the human prospects, the possibility of collaboration with other countries, the possibilities of control, control of information, the hope of an open attitude, above all the relations with Russia and what this would do to them, how it might injure or help them. I think that it probably was assumed, it certainly was always assumed at Los Alamos, that if the war were not over, and not clearly to be brought to a conclusion by diplomatic means, this weapon would play a part. I am not sure that the men who sat around that room all had the same idea of what would happen with the bomb and I therefore can neither confirm nor refute what Compton wrote. I think there was some discussion in which I didn't participate about the use of the bomb, but there was none in which I did.[6]

Also during the morning's session the future of fundamental nuclear research in the United States and its applications outside weaponry were discussed. Characteristically, Oppenheimer looked to the future of nuclear research as a Sesame for the good of mankind rather than being an omnipotent tool for destruction. Nonetheless, and Compton's memory singles out Oppenheimer as the physicist who provided the statistic, 20,000 was the figure quoted earlier in the morning as the number of deaths likely to result from the explosion of an atomic bomb over a city.

It was during the lunch break that, Arthur Compton recalled, some informal conversation began on what possibilities existed for not dropping the first bomb on a military target, but instead, turning it into a frightening display which would shock the Japanese into immediate submission. Compton was able here to introduce the attitudes of 'Christian compassion' he knew to be worrying some of his Chicago physicists and he sounded out Stimson's views on whether it was inevitable that many thousands of Japanese must necessarily be sacrificed in showing off the bomb's real power. There were several alternatives: the bomb could be dropped on an uninhabited island; it could be dropped on an uninhabited area of Japan, or Japanese observers could even be invited to a demonstration in the United States. Several of those at the table chipped in their views on the different possibilities.

But even if all those present were not fully aware of the new line the discussions had taken, the lunch-time group had too many obvious reservations to stay on this fresh tack for long. The first atomic bomb would be the *only* atomic bomb Los Alamos would have in a state of readiness for several days or even weeks; it might well turn out to be the most expensive fizzle-out in the history of fireworks. Even if a successful demonstration was carried out, it might convince nobody except the Americans themselves, and the element of surprise of the hard-kept secret would have been lost to no purpose. There were other fears. It was true that accusations of barbarism might follow hard after the first explosion on a Japanese city, but the manner of warfare of the enemy for whom the bomb was intended was not remarkable for its humanitarian qualities. Was it not possible that a Japanese government, warned of the first demonstration atomic explosion over its territory, would move into the area large numbers of Allied prisoners as a sacrifice to the weapon designed for its own destruction?

The Committee reconvened for the afternoon session, but the recommendations for the bomb's use were by now fixed. For some physicists who feared that rigid decisions of this kind might be being made in such places as the Pentagon where the whisper of the scientist might be drowned by the boomings of politicians and military men, the struggle to be heard now began in earnest.

One of the senior physical chemists who greatly feared that the bomb would be thoughtlessly and tragically thrown into the war before the scientists had organised themselves sufficiently to make their views known was yet another émigré: the Nobel prize-winner, James Franck. Franck, like other Jewish scientists, had joined the German army during the First World War, had been proud to associate himself with the rise of German militarism and had emerged with an Iron Cross and an unusually high field rank. A rich man, he personally helped finance the refitting of his laboratory in Göttingen and helped raise it to one of the most esteemed in the world. In the inter-war years Europe and America's young scientists – one of them Robert Oppenheimer – flocked there to fall under the influence of Franck's wisdom. But as with the rest of his Jewish colleagues, with the arrival of Hitler, wisdom, distinction in science, and even distinction in war, were found to count for nothing. Leaving his fortune in Germany and his Nobel Prize medal in Denmark (where it was dissolved in acid and recast after the war), Franck fled to the United States and eventually joined the Manhattan Project.

Just as Arthur Compton listened with great sympathy to the younger men, he listened with great respect to those who were of his own distinguished vintage. In particular he respected the views of James Franck.

Franck was one of a number of Chicago scientists to whom Arthur Compton spoke as soon as he returned from the vital Pentagon meeting. The workers at the Met Lab were now able to hear the good news that the Government was being kept advised on some of the problems which in the past weeks had been sending them into a flurry of activity. Oppenheimer, too, at Los Alamos, was telling some of his division leaders about the questions which were being raised in Washington. But it was from Chicago, the bubbling spring of discontent, that reaction to the news flowed.

Compton told his group that the Scientific Panel of the Interim Committee was due to meet in a few days time and that he would be willing to carry to it any recommendations on the future of

nuclear energy. Research physicists under normal working circumstances were not disposed to being committee men. But these
were not normal circumstances. The climax of the work of the
Chicago physicists was past, and the ends to which this work was
promised to be put were quite extraordinary. Within hours of
Compton's news no fewer than six committees had been formed
to give advice on Research, Education, Production, Controls,
Organisation and, eventually most significantly, Social and Political Implications. It was this last committee to which Franck acted
as chairman.

By 11th June the Franck Committee had completed its report.
Much of what the document contained had already been discussed
by concerned scientists in Chicago, Los Alamos, Washington and
other places in the United States. It dealt with the dangers of a
nuclear armaments race and the necessity of some form of international control. But also, and this qualified it for urgent consideration, a demonstration test of the bomb was recommended.
It said,

> . . . the military advantages and the saving of American lives
> achieved by the sudden use of atomic bombs against Japan may
> be outweighed by the ensuing loss of confidence and by a wave of
> horror and repulsion sweeping over the rest of the world and
> perhaps even dividing public opinion at home.
> *From this point of view, a demonstration of the new weapon might best be
> made, before the eyes of representatives of all the United Nations, on the
> desert or a barren island.* The best possible atmosphere for the
> achievement of an international agreement could be achieved if
> America could say to the world, 'You see what sort of a weapon
> we had but did not use. We are ready to renounce its use in the
> future if other nations join us in this renunciation and agree to the
> establishment of an efficient international control.'[7]

The members of the Franck Committee acknowledged that it
might well seem strange that scientists who had initiated the
development of the atomic bomb should be reluctant to try it out
on the enemy as soon as it was available, but they emphasised that
the compelling reason for creating the weapon had been a singularly forceful one: the fear of German technical skill and a
German Government which would have no moral restraints
regarding its use. A test demonstration which could take into
account the public opinion of other nations would mean the
possibility of sharing the responsibility for the fateful decision.

It was James Franck himself who took the report to Compton in the United States capital. Together they left it in Secretary of War Stimson's office.

By this document which embodies the ideas of many men who did not contribute directly to it, scientists had fully accepted their share of responsibility for their act of creation, and were, to the best of their abilities, attempting to warn mankind of its consequences.

The warning having been given and compliance having been made with the demands of conscience, all that was now required was to wait for the reaction. But no reaction came. Eugene Rabinowitch, who had drafted the Franck Report, was one of the ones sitting around and anxiously waiting with, as he later recalled, the dawning feeling that their action had turned into nothing more than that of a shipwrecked sailor who drops a desperate message in a leaking bottle into a great and stormy sea and hopes for a quick answer.

Only three days after Compton had met Franck in Washington, he travelled to Los Alamos. There, Oppenheimer had been asked to gather the Scientific Panel to consider among other things the question of a demonstration test. Though they might not have realised it at the time, the responsibility thrown on to the shoulders of these four men – Compton, Oppenheimer, Lawrence and Fermi – was huge. The consequences of the recommendations would shape the course of civilisation.

The conditions under which they had to meet were not easy. All four were still heavily involved in the preparations which would lead to the testing of the first weapon in the Alamogordo desert, now only weeks away. The tenor of the meeting, taking place as it did at the nerve centre of the whole project, could not fail to have been influenced by the hectic atmosphere of these last days of the crucial assembly, and the doubts and the fears of the scientists that were all around. Still no man could be sure that the device on the steel tower would perform any of the pyrotechnics that could be predicted on paper; whether it would splutter itself into ignominious failure was still a matter for conjecture. Even if it exploded with the release of the energy which Einstein's conversion of mass equation predicted, as a spectacle on a flat, barren desert, it could well fail to impress sceptical observers.

Compton recalls that it was Ernest Lawrence who was the last of the four men to give up finding some effective way of demonstrating the power of the bomb without loss of life. On 16th June they turned in their report to the Secretary of War. Stimson clearly interpreted one of the roles of the Scientific Panel as being one in which it could serve 'as a channel through which suggestions from other scientists working on the atomic project were forwarded to me and to the President'. In the Scientific Panel's recommendations he felt that he had taken the best advice available. In part these read:

> The opinions of our scientific colleagues on the initial use of these weapons are not unanimous: they range from the proposal of a purely technical demonstration to that of the military application best designed to induce surrender. Those who advocate a purely technical demonstration would wish to outlaw the use of atomic weapons, and have feared that if we use the weapons now our position in future negotiations will be prejudiced. Others emphasise the opportunity of saving American lives by immediate military use, and believe that such use will improve the international prospects, in that they are more concerned with the prevention of war than with the elimination of this special weapon. We find ourselves closer to these latter views; *we can propose no technical demonstration likely to bring an end to the war; we see no acceptable alternative to direct military use.*
> With regard to these general aspects of the use of atomic energy, it is clear that we, as scientific men, have no proprietary rights. It is true that we are among the few citizens who have had occasion to give thoughtful consideration to the problems during the past few years. We have, however, no claim to special competence in solving the political, social, and military problems which are presented by the advent of atomic power.[8]

The italics are Stimson's, used by him when he wrote in *Harper's Magazine* on 'The Decision to Use the Atomic Bomb' almost two years after the recommendations were made. They seem a clear admission of how strongly these phrases of the report influenced him, even though in the same paragraph in which they were contained it was pointed out that scientific opinion was divided on the issue. Oppenheimer was never able to recall with any sureness the date on which he read the full contents of the Franck report though it was probably after the Scientific Panel's meeting. When he did so he found it a moving, eloquent and thoughtful document. At the time of the meeting, however, under the influence of Niels Bohr's reasoning, the emphasis of Oppenheimer's concern was with the political problems and future relations with Russia, rather than what one would do with the bomb once it was ready. Arthur Compton, nonetheless, was fully aware of what the Franck document contained and must have brought its main points, if not the document itself, to the attention of the other three physicists. General Groves, the military head of Los Alamos, recollects having informed the Interim Committee of the contents of the Franck Report, and that it was carefully considered by the Scientific Panel. In spirit, if not in actual fact, the Franck report was rejected. Justice had held out the scales of right and wrong on which the scientists could load their weights and the heaviest were on that pan which said use the bomb.

There was one scientist who would never have been daunted by the possibility of influencing single-handed a figure as imposing as that of Justice herself. In his peripatetic fashion Leo Szilard had, in the past, tackled, or attempted to tackle, perhaps equally imposing figures. In May of 1945 he had tried to give his views on nuclear energy to the President of the United States, but had had to content himself with a visit to Truman's adviser, James Byrnes, at Byrnes' home in South Carolina. The Southerner had found the Hungarian's manner not at all to his liking (even some scientists found Szilard's behaviour a little odd on occasions), and Szilard came to the conclusion (as did other scientists) that Byrnes did not understand the full implications of nuclear energy.

This disastrous rebound of personalities only encouraged Szilard into more beavering activity. He had been a member of the Franck Committee and, in common with Franck, was particularly concerned with how the bomb would be used. He was aware that a test of the weapon was imminent and so began to draw up a petition to the President. The petition went through many phases before Szilard managed to collect almost 70 signatures, including those of some of the leading atomic scientists, to a document which pointed out that

> ... a nation which sets the precedent of using these newly liberated forces of nature for purposes of destruction may have to bear the responsibility of opening the door to an era of devastation on an unimaginable scale.[9]

It petitioned the President not to use the atomic bomb unless Japan refused to surrender to publicly declared conditions. It ended by encouraging Truman to consider all the grave moral responsibilities involved before deciding whether or not to use the bomb.

Szilard had collected most of his signatures by circulating his petition in Chicago; in other laboratories he had had surprisingly little success. In an attempt to give some support from Los Alamos, Szilard wrote to his fellow Hungarian, Edward Teller, asking for his blessing and suggesting that he get others to join. Teller, who approved both Szilard's petition and methods, before putting the document on its rounds, decided first to get the approval of his director, Robert Oppenheimer.

It was, according to Teller, his duty to take this petition to Oppenheimer; but more, Oppenheimer, the natural leader, 'seemed to be the obvious man to turn to with any formidable problem, particularly political'.

Memory is sharply divided on the incident. According to Teller, Oppenheimer 'thought it improper for a scientist to use his prestige as a platform for political pronouncement'[10] and lifted a great weight from Teller's heart by assuring him that the best men Washington could find were applying their wisdom to the decision on the manner in which the bomb would be used.

Oppenheimer, 20 years after the event, had no memory of any such discussion, and though prepared to admit that it is possible that he might have known of the Szilard document when it was circulated, he believed that he never found out about it until after the War.[11]

Today Edward Teller regrets not having circulated the Szilard petition at Los Alamos; he took the advice which he remembers Oppenheimer offered him 'in no uncertain terms', not to participate.[12] He also took the advice not to involve anybody else, and apparently to good effect: very few Los Alamos workers besides Teller can remember having heard of Szilard's petition at that time. Even Robert Wilson, worried as he was by the impending impact of nuclear weapons, and who as a Los Alamos official travelled back and forth between Chicago and Los Alamos, got no wind of what Szilard was up to. But when news of the goings on in the north filtered down to the New Mexico laboratory it produced no stir and no stimulus to any further action. There was a great deal else to be concerned with and there were many others besides Teller who felt that the wisdom and deep concern of

'Oppy' the leader were capable of making judgments and decisions on grave issues to which they themselves could give little time and thought.

The date on which their thoughts were focused was 16th July 1945. That was the day on which the explosion in the Alamagordo desert left no doubts about the devastation of which an atomic arm was capable. The final version of Szilard's petition was dated 17th July. It was to Arthur Compton whom Szilard gave his document and its collection of signatures for delivery in Washington.

Compton, perhaps more than any other physicist, was uncomfortably aware of a great diversity of opinion in the Chicago moral hot-spot. Szilard's had been by no means the only piece of paper which Compton knew to have been moving around offices and laboratories gathering autographs. For example, several Chicago scientists had signed a petition agreeing generally with Szilard, but being particularly concerned with the use of the bomb against cities. Another large group in Oak Ridge had also petitioned that the powers of the weapon be demonstrated and that the Japanese should be made aware of the nature of the catastrophe to which refusal to surrender might lead.

Arthur Compton had already given his own mark of approval to the recommendation from the Scientific Panel. The knowledge that among many of his scientists there was unequivocal opposition to this view must have troubled him deeply. In his memoirs of this period he goes to some length to point out the strength of scientific support for the decision he had already made. He quotes what he describes as 'a counterpetition' reading,

> . . . are we to go on shedding American blood when we have available a means to speedy victory? No! If we can save even a handful of American lives, then let us use this weapon – now![13]

In detail he describes a young physicist who visited his Chicago office with tears in his eyes as he pleaded that every available weapon be used for the benefit of his buddies at that moment being wounded and killed on the Far Eastern battlefields.

Perhaps hoping to demonstrate to others some endorsement of the decision to which he had already committed himself, Compton agreed to supervise an opinion poll among the Chicago scientists who were familiar with the nature of the bomb and the deep issues involved. The poll, taken in a secret ballot on 12th July, was a voluntary one among physicists, chemists, biologists, and metallurgists who had academic degrees; it reached 150 scientists: about half the Met Lab staff.

Several possible procedures were put to the vote; that which produced by far the greatest response, 46 per cent, was in favour of 'a military demonstration in Japan to be followed by renewed opportunity for surrender before full use of the weapon is employed'.

On 24th July Colonel Kenneth Nichols, the 'Manhattan District' Engineer, went to Compton at Oak Ridge and asked him for the results of the opinion polls. Compton wrote out a letter pointing out which procedure, according to his own interpretation, was most favoured. An hour later Nichols was back in Compton's office saying, 'Washington wants to know what *you* think'. For Compton, who had had to face that question many times during the past weeks, this was the crucial time of asking, even though he had already answered. At that moment he still believed his influence to be such that 'a firm negative stand on my part might still prevent an atomic attack on Japan'. His mind churned through the conflicting motives which could influence his reply: his pacifist family tradition, the suffering he knew the weapon must cause, his wish to see the war at a quick end and the lives of the many young Americans he knew which would be saved if such a quick end were achieved. His reply to Nichols was, 'My vote is with the majority. It seems to me that as the war stands the bomb should be used. . . .'[14]

But what *did* he think? And what did the majority think? Compton's own decision had already been stated to the Secretary of War along with that of Oppenheimer, Fermi and Lawrence; he had seen 'no acceptable alternative to direct military use'. But, clearly, some of the scientists who subscribed to the 'majority' poll recommendation which he had just interpreted for Washington's benefit might well have interpreted the phrase 'a military demonstration in Japan' as a demonstration test with a forewarning, whereas Compton had apparently interpreted it as an attack without warning.

Even if Compton, a man who had racked his conscience more than most over the issues involved, had misinterpreted the results of a badly worded referendum, did it in the least matter? Could his or any other scientist's influence affect the bomb's use? Colonel Nichols sealed up in a brown paper package, addressed to General Groves in Washington, all the documents intended for the highest authority of the United States Government – the Met Lab poll, Szilard's petition and several other petitions and letters from scientists; they reached Washington next day. And if this confusing mixture of conviction could have been satisfactorily interpreted, neither Truman nor his Secretary of War were around the capital to receive it. They were in Europe at the Potsdam meeting with Stalin and Churchill.

While the brown paper package was on its way to Washington, the civilian-suited Truman had wandered round the Potsdam conference table to the uniformed Stalin where, with deliberate casualness, he had passed on the news that the United States now owned a new weapon of great destructive power. He did not mention that it was nuclear. If the news was of any special interest, the moustached face above the uniform did not register it as such: perhaps because Russian Intelligence had been quicker with the information than had Truman. Whatever the source of the phlegm, Stalin wished Truman well in the use he could make of the innovation against the Japanese.

On the next day Henry Stimson finally approved the military operation which would deposit the first of the new weapons on Japan. The bomb was to be delivered as soon after 3rd August as weather conditions would permit, and the target was to be one of four Japanese cities: Hiroshima, Kokura, Nagasaki or Niigata. Stimson, sensitive to the reaction the use of the bomb might bring, had had removed from the list the old capital of Kyoto, a centre of Japanese culture and art, which was originally chosen by a target committee operating under the direction of General Groves.

On 26th July the Potsdam Proclamation to the Japanese Government was broadcast across the Pacific from the West Coast of America. On behalf of their peoples, President Truman, President Chiang Kai-shek and Prime Minister Churchill called for the unconditional surrender of all Japanese armed forces. Should the Japanese government fail to comply with this ultimatum, then it was promised that 'the full application of our military power, backed by our resolve, will mean the inevitable and complete destruction of the Japanese armed forces and just as inevitably the utter devastation of the Japanese homeland'. No indication was given of how the Allied leaders intended to implement this utter devastation.

Two days passed before Japan's Premier, Suzuki, rejected the ultimatum as unworthy of public notice, and Radio Tokyo told of the proud attitude to the Japanese people.

By 1st August Stimson was back in Washington and it was to his office that General Groves delivered the bundle of petitions Nichols had sent him from Oak Ridge. But whatever decision these pleas might have hoped to have influenced had already been made. The package was carefully filed.

So it was that, whilst physicists were still discussing, petitioning and voting in an attempt to have some consideration given to their views of how their scientific creation should be employed in warfare, and whilst a few – Arthur Compton was one – were still of the belief that a word by them in the right place would influence the decision to make an atomic attack on Japan, final details on the use of the bomb were being rounded off in Potsdam and in Washington.

There is no shortage of men who believe that their words clinched whatever decision had to be made on the use of the first nuclear weapon. On the choice of targets Stimson was convinced that the decision was his alone and at an early stage had told General Groves, 'This is one time that I'm going to be the final deciding authority. Nobody's going to tell me what to do on this. On this matter I am the kingpin.' On the use of the weapon without any prior technical demonstration to the Japanese he wrote, 'The ultimate responsibility for the recommendation to the President rested on upon me, and I have no desire to veil it.' And Truman also wanted no disclaimers and wrote,

> The final decision of where and when to use the atomic bomb was up to me. Let there be no mistake about it. I regarded the bomb as a military weapon and never had any doubt that it should be used.[15]

But it is more than doubtful whether one man alone, be he scientist or politician, was ever capable of stopping the roll of the technological ball which, once started on its steep and costly path to the nuclear weapon, quickly developed an overpowering momentum. The history of the atomic project is apparently that of an inevitability which swept over the scientists whose protests, when they were made, were made too late even to deviate the thing on its downward track. Churchill demonstrated how little the influence of the working physicist managed to percolate through to his political stratum, when he wrote of Potsdam,

> ... the decision whether or not to use the atomic bomb to compel the surrender of Japan was never even an issue. There was unanimous, automatic, unquestioned agreement around our table; nor did I ever hear the slightest suggestion that we should do otherwise.[16]

British approval for the use of the bomb had been sought and given, but Churchill was not to be leader of the nation when the Japanese Premier rejected the Potsdam ultimatum. The General Election of July 1945 swept Churchill out and Attlee into office. So it was that, when Premier Suzuki, on behalf of his country, refused to surrender and thus made the use of the nuclear weapon inevitable, both the United States and Britain were being led by newcomers to the idea of an atomic bomb. Truman had known of it for only three months, whilst Attlee, even at this time, knew no more than that it was an explosive more powerful than any other yet used.

In the early hours of 6th August, a B-29 bomber left the island of Tinian, in the Marianas in the Pacific Ocean, to begin its journey of about 1,500 miles north to Hiroshima. Its pilot was Colonel Paul Tibbets who had named the plane after his mother, Enola Gay, so presenting her with a perhaps unwelcome immortality.

The 'Enola Gay' was carrying 'Little Boy': the code-name for a bomb of the uranium gun type with a power greater than 20,000 tons of TNT. One group of Los Alamos scientists had helped load 'Little Boy' and a second group was travelling in one of two B-29 observer planes in order to measure physical effects such as the yield of the bomb.

Weather conditions were clear when the planes approached the target area. Just before the bomb was due to be dropped the physicists left the small window through which they were watching the approaching Japanese city and returned to their measuring devices which were to be parachuted to the area below. The 'Enola Gay' released its package and banked sharply. In the bright August morning, the atom bomb fell close on target.

When the explosion came the physicists were working at their instruments. The inside of the observer plane lit up as though a flash bulb had been let off there. There were seconds of worried waiting for the blast to strike the fuselage; when the buffeting and the rattle had stopped, they could again crowd to the little window. One of them, Harold Agnew, took out his 16 mm. camera and made the only film record of the rise of the tall mushroom cloud above the pall of smoke which was the ruined city.

At Los Alamos itself Otto Frisch heard the announcement of the successful explosion of the Hiroshima weapon over the internal public address system. It was read by a girl operator who clearly did not understand the significance of her message. The reaction of the laboratory's scientists to the news they had been expecting for some days was a very mixed one. Frisch could hear around him the thump of running feet in the corridors of the prefabricated buildings and the yells of some of his colleagues calling to one another and rushing to the telephone to order celebration dinners at nearby restaurants. He himself felt sick at heart that the diners would be celebrating, not just a technological success, but the deaths of many thousands of people.

Richard Feynman, also hearing the news over the tannoy, joined in the general euphoria he found spreading in the men around him. But, for all his excitement, he became conscious of a figure he knew well standing out from the group and leaning dejectedly against a wall. It was Robert Wilson, the physicist who at Princeton had persuaded Feynman to leave his Ph.D. thesis and join the bomb project, and who at Los Alamos had organised the 'Impact of the Gadget' meeting. Feynman walked over to find out the cause of the dejection to be told by Wilson that he was desperately concerned by the serious news of the bomb being put to use. By contrast with the mood of gaiety and laughter spreading around them, Wilson's attitude was paradoxical and seemed almost funny. Feynman recalls saying, 'Listen, you started the whole thing; you got me into it. Cut it out. You know you were trying to do it all the time.' Feynman ran off laughing. Not until later did he begin to appreciate the reasons for the depth of Wilson's concern.

On the day that a fifth of the city of Hiroshima's 300,000 inhabitants were killed by the explosive effects, the flash burns and the radiation of the uranium bomb, Leo Szilard requested permission to publish his petition to the President of the United States. In the days during which he waited for a reply, the enemy nation was assessing the extent of the disaster. Now the Japanese Government knew the meaning of the Potsdam ultimatum and the 'utter devastation' it had promised.

On 9th August Szilard was told that his request had been disapproved. That day a number of physicists helped load 'Fat Man', a plutonium implosion weapon, on to Major Charles Sweeney's plane 'The Great Artiste'. Three of the scientists attached to some of the instruments to be dropped at the same time as the bomb, a message hand-written by Luis Alvarez and addressed to Professor R. Sagane: a Japanese scientist who before the war had worked with them in America. In this, they asked him, as a nuclear physicist, to use his influence on the Japanese leaders to prevent any further suffering. It ended, imploring '. . . do your utmost to stop the destruction and waste of life which can only result in the total annihilation of all your cities if continued. As scientists, we deplore the use to which a beautiful discovery has been put, but we can assure you that unless Japan surrenders at once, this rain of atomic bombs will increase manyfold in fury'.[17] It was signed, 'From: Three of your former scientific colleagues during your stay in the United States'.

News of the explosion of the Nagasaki bomb reached the Japanese Cabinet in a late-night sitting when it was still in disagreement about what its attitude should be. By early next morning Japan had surrendered.

That day Eugene Rabinowitch walked into the Metallurgical Laboratory in Chicago and saw from the faces of the research workers there that something was wrong. The news was broken to him that Nagasaki had been devastated, and like the rest who had signed the Franck report, and like the others who had tried to plead for a demonstration test or for some deliberation of the moral considerations, he was shocked and bitter.

At the time of the dropping of the first bomb on Hiroshima scientists were completely divided on their views of the use of this first atomic weapon. In the intervening years some have changed their attitudes or formed new ones which on 6th August 1945 could not have existed.

Edward Teller's attitude, in the main, has not changed since he found himself in agreement with Szilard's petition. It is,

> In my view it was wrong. If we had made a demonstration and that had failed, then I think dropping the bomb would have been justified in order to end the war. To drop it without warning was wrong. It was wrong on moral grounds – it killed; it was wrong, although I could not see that at the time, on practical grounds because the dropping of the bomb has distorted our views, has changed our whole outlook. We are not now looking on the accomplishment of atomic explosions as progress which can, and should, be used in the right way. We had started at that time to look at it as something horrible, something that should not be continued.[18]

Robert Wilson, though willing to acknowledge that there were good reasons at the time to make the decision and to bring the war to as rapid an end as possible, holds a similar view:

> I think that we as a nation will go down in long history as the first nation to have used atomic energy for this purpose. This will just be a blot that we will carry through all historical times. I feel it was wrong.[19]

Many scientists hold like opinions. But there are others who hold the contrary view. For example, Harold Agnew, the young physicist who flew with the Hiroshima mission:

> I think it was right to drop the bombs because I believe that the dropping of these bombs brought the war to a close much quicker than would have been possible otherwise. I think if people who now debate this question had seen the preparations which we were making for evacuating the wounded, the hospital preparations and everything, anticipating an actual landing, that they would have realised that we actually saved lives: not only our own soldiers' lives, but the lives of the Japanese, because had we been forced to actually attempt to occupy the island I think the death toll would have been tremendous.[20]

There are many others, particularly those with experience of the results of the bombing raids of Europe, such as Otto Frisch and Jim Tuck, who, like Agnew, believe that the overall effect of the devastation might have been an economy of lives. But history is such that there is no method of verifying these retrospective conjectures. As Robert Oppenheimer has said, not even the experts of warfare have any means of knowing whether the war could have been terminated without the atomic weapon:

There are passionate arguments: they do not persuade me one way or the other. At the time the alternative campaign of invasion was certainly much more terrible for everyone concerned. I think that Hiroshima was far more costly in life and suffering and inhumane than it need have been to have been an effective argument for ending the war. This is easy to say after the fact.[21]

For after the fact it becomes too late to condemn meaningfully those who contributed to the decision to use the first atomic weapon in warfare without warning. The facts which had to be taken into account to reach this decision were those which were known during the first few days of August of 1945, and not those which are known to us now in the light of history. At the time of Hiroshima (as at the time of Nagasaki) only one bomb was available and no man had any sure way of knowing whether it would explode. The same situation would have held if a test demonstration had been attempted. Though it was suspected that Japan was on the very brink of surrender, there was again no sure way of knowing if and when she would do so, and meanwhile the toll of life in the continuing war would mount. In terms of both American troops and Japanese soldiers and civilians the death roll, which must have resulted from an invasion of the Japanese islands as the only alternative to the use of the bomb in the continuing war, could only have been enormous.

The reaction to the suffering inflicted on the inhabitants of Hiroshima and Nagasaki was inevitable. What surprised scientists, however, was not that this reaction should have set in at all in a war of protracted suffering and horror, but that the finger of guilt should have been pointed at science. To the world at large, physics, by its very ability to provide the means of creating the nuclear weapon, was seen by some eyes as being as responsible as the political machine, or the war machine, which caused it to be put to use. It was what many people wanted to hear when Robert Oppenheimer said of his and his fellow scientists in the atomic bomb project, 'the physicists have known sin'. By this he meant,

> We were all aware of the fact that in one way or another we were intervening explicitly and heavy-handedly in the course of human history; that is not for a physicist a natural professional activity.[22]

But given the face of warfare and the facts that went along with it in 1945, history will always show that there existed a group of scientists, and Robert Oppenheimer was one of these, who were deeply aware of the implications of their work and of the necessity of preparing the world for the time when it would come to fruition; and that there existed some who, before the event, so strongly disagreed with the way in which the atomic bomb was used, that they attempted to influence the decision by drawing to the attention of the political leaders responsible for it, the consequences mankind might have to face. Almost all these men were scientists who had contributed to this weapon of war and were attempting to face the responsibilities which they believed their part in the act of creation had given them. That their actions came too late, or that they were ineffective, is perhaps less important in the long run of civilisation than that they should have acted at all.

On 16th September 1967 a young Russian physicist was reported to have cried for help as he was dragged into a car standing on the Bayswater Road. A few hours later he was seen acting as the human rope in a tug-of-war between British police and members of the crew of a Moscow-bound Russian TU 104 jet on an airstrip at Heathrow. The news, not unnaturally, reached world press headlines and the incident caused tremors at the highest British and Soviet governmental level. But the situation was scarcely a novel one. *The Times*, reporting the affair, had no difficulties in listing precedents. In 1954 the wife of Vladimir Petrov, third secretary at the Russian Embassy in Canberra, had left an aircraft at Darwin after a crowd of many hundreds had struggled with Russian officials at Sidney airport, when Mrs Petrov had been reported as crying out, 'I don't want to go. Save me.' In 1961 the ballet dancer, Rudolf Nureyev, had run from a group of Russian security officers at Le Bourget airport where they were trying to put him on an aircraft to Moscow; he had been heard to shout, 'These men are kidnapping me. I want French protection. I want to be free.' In 1963 a Russian woman schoolteacher in Dakar had been freed by police after Russian Embassy officials had tried to put her on a Moscow-bound flight with a false passport. There are other examples.

The young physicist, whose name in 1967 was added to the list of Russians apparently being forced to return to their homeland against their will, was Dr Vladimir Kachenko, an exchange post-graduate student doing low-temperature research work at Birmingham University. And what, to newspapers in the few hours after the incident, made his case particularly outstanding in this long run of aborted abductions, was, not so much the fact of its occurrence, but more the possibility that the immigration officers who prevented his flight to Moscow had erred badly; there seemed a chance that the Soviet Embassy was moving the young man at speed for the best humanitarian reasons. Indeed, within 48 hours the British authorities had, so it seemed, acknowledged their mistake by handing back Kachenko, apparently in a serious mentally disturbed condition, to his Embassy for treatment in his home surroundings, which was what the Embassy claimed it was attempting to give him in the first place.

But there was once a time when incidents involving unwilling Russian repatriates were not accepted as commonplace; none of the overt diplomatic squabbles between Russia and nations trying to protect Russian subjects have had such a profound and shocking effect on the Western nations as did the very first in line of these affairs. It occurred so shortly after the end of war in 1945, so shortly after the wartime propaganda machine had ceased engineering a smooth surface on all the faces of the Russo-Anglo-American grand alliance, as to make those first involved in it refuse to believe in the cries for help they could hear. For they had had no precedent to make them over-eagerly dash to the aid of a Russian apparently trying to defect. A quarter of a century ago the word defection had no meaning as we know it now.

On the evening of 24th September 1945, Igor Gouzenko, a 26-year-old cipher clerk on the Soviet Embassy staff in Ottawa, left his office carrying a bunch of papers. He was a frightened man and there was good reason for his fear. During the past several weeks he had been noting carefully a selected number of the documents of his Military Attaché, Colonel Zabotin, by turning down the corners of pages he knew to be of special interest. He had not, however, moved them from their files. Then, just before 8 p.m. that evening, he had collected all these papers and started out from the Embassy. He headed immediately for one of Ottawa's daily newspaper offices. But here the plan he had been making with such deliberation over those last weeks, even now as he was launching it, began to his great peril to go wrong. He could find nobody in the whole of the newspaper office to take him seriously, nor to believe that the documents he was carrying were genuine. It was a chilling situation which, in his best laid scheme, he had understandably failed to reckon for.

During the earlier part of the next day he again tried to tell his story at the newspaper office; again he failed. Then he tried various government offices with the same mountingly distressing result. Whether because of the apparent exaggeration of the sensational nature of what he had to tell he was assumed to be a trickster after cash, or whether because of his stage Russian accent he was assumed to be a joker or a lunatic, he could nowhere make anybody believe that he was what he said he was: a defector. (Twenty-two years later the political climate had so reversed itself that London airport immigration officers could not believe that Dr Vladimir Kachenko was mentally ill and that he was not a defector.)

It was about 7 o'clock in the evening of 6th September that a Non-Commissioned Officer of the Royal Canadian Air Force, sitting with his wife and children on the balcony of his Ottawa apartment, saw on the adjoining balcony his neighbour, Igor Gouzenko, beckoning to speak to him. Describing the incident some months later, the N.C.O. said,

> I told him sure he could speak with me, if he had something to say; so he asked me if the wife and I would look after their little boy if anything should happen to him and his wife. So about that time I figured maybe we should go inside, so we went into our apartment, and while in there he said he figured that the Russians were going to try to kill him and his wife, and that he wanted to be sure that somebody would look after his little boy if anything should happen to them.
> So after a bit of a conference my wife and I decided we should look after him, because we didn't want to see him stuck with nobody to look after him should anything happen to them.[1]

To what must have been his everlasting surprise, by his perhaps unsophisticatedly ready acceptance of this unlikely tale, the N.C.O. began the first stage in the exposure of a nation-wide Fifth Column, and made possible the first of a long series of revelatory defections.

Perhaps at that stage of the evening the airman in no way believed the story that murder was likely to be committed in the next apartment, but no matter: Gouzenko raised enough compassion to have his son taken in. Before the night of the 6th–7th was out, however, the N.C.O. knew for sure that something strange was threatening the once quiet life of Gouzenko. The Russian became so upset by the sight of a figure prowling in the lane behind the apartment that the N.C.O. allowed himself to be persuaded into looking after, not only the child, but both parents as well.

Between 11.30 and midnight four men, obviously also Russians, appeared in the building and began knocking on the door of Gouzenko's apartment, number 4. The airman spoke to them from the door of number 5 and, in answer to their questions, told them he knew nothing of Gouzenko's whereabouts. The four men now went downstairs as if to leave but, as could be plainly enough heard from next door, returned within a short time, broke the lock on Gouzenko's door and began ransacking the apartment.

At this point the police, who had been alerted earlier in the evening by a quick bicycle sortie of the N.C.O., stepped into the middle of the bizarre scene and so ensured Gouzenko the publicity he needed for his safety. If he had something worthwhile to tell, then it would now be listened to.

Gouzenko had most certainly got something to tell, and the four members of the Russian Embassy staff who, in the early hours of 7th September, angry and empty-handed, left apartment 4 under the perplexed gaze of a pair of Canadian police constables, had failed in their mission to prevent him telling it.

Gouzenko spent the rest of the night in the apartment block under police care, and when day came was taken to a Royal Canadian Mounted Police office. There he turned over some documents which were to put on display a great espionage net: a Fifth Column organised and directed by Russian agents in Canada and in Russia, into several interlocking spy rings. Canadian citizens were deeply involved, as well as the citizens of other Western nations whose compatriots, until the Gouzenko affair was made public, had little idea of the true relationships existing between Russia and her war-time Allies.

And caught in this mesh was a scientist: unambiguously implicated by Gouzenko's document. His case was to make other scientists look searchingly at yet another aspect of the responsibilities which weapon research necessarily thrust into their territory. This man found himself with an acute problem of conscience which he chose to solve in a fashion he knew to be punishable by law. He was Alan Nunn May.

It was in the spring of 1942 that May, an experimental physicist at King's College, London, was recruited to the uranium research work under the 'Directorate of Tube Alloys' (the cover-name for the British atomic project). In January 1943 he travelled to Canada to join the Anglo-Canadian atomic project and soon immersed himself in the work going on at the Montreal Laboratory, where he became a group leader, and at Chalk River where a pilot nuclear pile was being built. In Canada he became a trusted member of a hard-worked international scientific outfit staffed by his own countrymen, Canadians, New Zealanders, Frenchmen and some European refugees; his knowledge, however, was not restricted only to the Canadian aspects of progress towards nuclear power. The Americans were providing a generous amount of material – heavy water and samples of separated uranium isotopes – and also much information. And from his base in Montreal, May was able to supplement his knowledge with a number of visits to the Metallurgical Laboratory in Chicago. But though May's colleagues in Canada had no suspicions of his behaviour during his stay in that country, not so some of the Americans. Arthur Compton on one occasion became so apprehensive of some of the detailed questions which May was beginning to ask, that he told him that he did not consider it to be in the interest either of the development of the U.S. or the Canadian programme to give him the information he was after. When May persisted, Compton found it necessary to tell him he was no longer welcome at the Met Lab.

There was, however, no suspicion anywhere in the Allied project that May was what he was: a most zealous but undeclared Communist with highly personal views of where his duty lay. Already well known to the right authorities in Moscow, he was ideally placed as a source of atomic secrets. Soon after his arrival in Canada he was contacted by a member of Colonel Zabotin's organisation at the Soviet Embassy in Ottawa and given the cover-name 'Alek'. And when in the summer of 1945 news reached these authorities that his tour in Canada was coming to an end, the wires between Moscow and Ottawa began to demand the milking of May to the limits of his knowledge on atomic matters. The documents which Gouzenko had carried out of his Embassy cipher office gave full details of these demands and how they had been answered. They were later to be published in full in a report by a Canadian Royal Commission on espionage.

This report shows that, on 28th July 1945, 'The Director' in Moscow telegrammed to 'Grant' – the cover-name for Colonel Zabotin – in Ottawa,

> ... Try to get from him before departure detailed information on the progress of the work on uranium. Discuss with him: does he think it expedient for our undertaking to stay on the spot; will he be able to do that or is it more useful for him and necessary to depart for London?

On 9th August, the day the plutonium implosion weapon was dropped on Nagasaki, Zabotin replied to Moscow,

> Facts given by Alek: (1) The test of the atomic bomb was conducted in New Mexico (with '49', '94–239') [i.e. *plutonium*]. The bomb dropped on Japan was made of uranium 235. It is known that the output of uranium 235 amounts to 400 grams daily at the magnetic separation plant at Clinton. The output of '49' is likely two times greater (some graphite units are planned for 250 megawatts, i.e. 250 grams each day). The scientific research work in this field is scheduled to be published, but without the technical details. The Americans already have a published book on this subject.
> (2) Alek handed over to us a platinum with 162 micrograms of uranium 233 in the form of oxide in a thin lamina. We have had no news about the mail.

May had done better than merely handing over 'detailed information on the progress of the work on uranium'; he had somehow managed to get hold of uranium samples which he gave to Lieutenant Angelov, one of the Secretaries of the Military Attaché. And so important were these samples considered to be that a second officer, Lieutenant Colonel Motinov, was despatched with them to Moscow.

A few days after this most unexpectedly fruitful meeting of May with his staffman, Zabotin himself took off to see what he could see. He paid a call on a friend who lived in the neighbourhood of Chalk River, and used the visit to take a brief river cruise from which he could observe the atomic plant. He reported to Moscow with the details.

May, who had still two more calls to pay on Chalk River, was by now nearing his departure date and 'The Director' was keen to maintain contact in Britain. Zabotin therefore was sent the following telegram from Moscow in late August:

> Work out and telegraph arrangements for the meeting and the password of Alek with our man in London.

Zabotin replied saying that May was to work at King's College, Strand, and that it would be possible to find him there through the telephone book. Three possible dates in October for meetings in front of the British Museum were proposed, along with an identification sign and a password. 'The Director', however, was not best pleased with Zabotin's logistics and replied:

> The arrangements worked out for the meeting are not satisfactory. I am informing you of new ones.
> 1. Place: In front of the British Museum in London, on Great Russell Street, at the opposite side of the street, about Museum Street, from the side of Tottenham Court Road repeat Tottenham Court Road, Alek walks from Tottenham Court Road, the contact man from the opposite side – Southampton Row.
> 2. Time: As indicated by you, however it would be more expedient to carry out the meeting at 20 o'clock, if it should be convenient to Alek, as at 23 o'clock it is too dark. As for the time, agree about it with Alek and communicate the decision to me. In case the meeting should not take place in October, the time and day will be repeated in the following months.
> 3. Identification signs: Alek will have under his left arm the newspaper *Times*, the contact man will have in his left hand the magazine *Picture Post*.
> 4. The Password: The contact man: 'What is the shortest way to the Strand?'
> Alek: 'Well, come along. I am going that way.'
> In the beginning of the business conversation Aleck says: 'Best regards from Mikel.'
> Report on transmitting the conditions to Alek.

May visited Chalk River on 16th August, and then for the last time on 3rd September. Two weeks later he was back in England. But neither in October nor in any of the following months did he put into operation in Great Russell Street any of the intricate recognition procedure which the author of a spy novel might well have deleted from his plot as being too melodramatic for reality.

On 6th September Gouzenko had appeared on the balcony of his apartment to attract the attention of his airman neighbour, that of the Canadian Government, and of course that of 'The

Director' in Moscow. Known or unknown to May his movements from a time shortly after he returned to England were under the close scrutiny of the Special Branch of Scotland Yard.

At about 3 o'clock in the afternoon of 4th March, 1946, the recently appointed Reader in Physics at King's College finished a lecture he was giving to a small group of his students, then went off in a taxi with two detectives to Bow Street where he was charged under the Official Secrets Act. There, at a hearing two weeks later, before a crowded court, the Director of Public Prosecutions, Mr Anthony Hawke, told the Magistrate some of the admissions which this physicist, Dr Allan Nunn May, had made in a signed statement. It had read:

> About a year ago, whilst in Canada, I was contacted by an individual whose identity I decline to divulge. He called on me at my private apartment in Swail Avenue, Montreal. He apparently knew I was employed by the Montreal laboratory and he sought information from me concerning atomic research.
> I gave and had given very careful consideration to correctness of making sure that development of atomic energy was not confined to U.S.A. I took the very painful decision that it was necessary to convey general information on atomic energy and make sure it was taken seriously. For this reason I decided to entertain a proposition made to me by the individual who called on me.
> After this preliminary meeting I met the individual on several subsequent occasions whilst in Canada. He made specific requests for information which were just nonsense to me – I mean by this that they were difficult for me to comprehend. But he did request samples of uranium from me and information generally on atomic energy.
> At one meeting I gave the man microscopic amounts of U233 and U235 (one of each). The U235 was a slightly enriched sample and was in a small glass tube and consisted of about a milligram of oxide. The U233 was about a tenth of a milligram and was a very thin deposit on a platinum foil and was wrapped in a piece of paper.
> I also gave the man a written report on atomic research as known to me. This information was mostly of a character which has since been published or is about to be published.
> The man also asked me for information about the U.S. electronically controlled AA shells. I knew very little about these and so could give only very little information.
> He also asked me for introductions to people employed in the laboratory including a man called Veall but I advised him against contacting him.
> The man gave me [*the figures '200 ANM' had been written in here,*

then crossed out] some dollars (I forget how many) in a bottle of whisky and I accepted these against my will.

Before I left Canada it was arranged that on my return to London It was to keep an appointment with somebody I did not know. I was given precise details as to making contact but I forget them now. I did not keep the appointment because I had decided that this clandestine procedure was no longer appropriate in view of the official release of information and the possibility of satisfactory international control of atomic energy.

The whole affair was extremely painful to me and I only embarked on it because I felt this was a contribution I could make to the safety of mankind. I certainly did not do it for gain.

At the hearing Mr Hawke told the court that, before accepting his position as a temporary civil servant, May had been required to sign a document addressed to the Department of Scientific and Industrial Research in which he undertook to observe secrecy in his work in connection with 'Tube Alloys'. When he had gone to Canada it had again been brought to his attention that the same conditions of secrecy and confidence were imposed on him. He had accepted these conditions.

Giving evidence, Deputy Commander Leonard Burt of the Special Branch of Scotland Yard, who also held the rank of lieutenant colonel on the Intelligence Staff, described how he had interviewed May on a number of occasions during February, and the circumstances under which the prisoner had signed his statement. Burt, after a question from Mr Gerald Gardiner, defending counsel, agreed that Dr May was a scientist of repute and had shown an exemplary character.

Also appearing to give evidence was Sir Wallace Akers, member of the board of Imperial Chemical Industries and Director of Tube Alloys. During questioning from Mr Gardiner there was a curious interchange which was reported as:

Mr Gardiner: Is there strong feeling among many scientists, rightly or wrongly, that contributions to knowledge made by them with respect to the benefits of atomic research ought not to be the secrets of any one country?
Sir Wallace: There is a strong feeling.
Mr Gardiner: In February last year Russia was a gallant ally, was she not?
Sir Wallace was understood to say 'Yes', but the magistrate intervened, saying: 'Surely it is a matter of historical record and not for the depositions. It is largely a matter of opinion on which a

chemist is not necessarily expert. I do not think it is a question Sir Wallace can be asked. I think we can leave it out.'

Mr Hawke: I would like to know why Russia has been introduced at all. I have made no mention of Russia or America or anybody. [*No mention had yet been made of the nationality of Nunn May's contact.*]

Mr McKenna (The Magistrate): I shall rule it out.

Mr Gardiner: I asked it quite deliberately, and if you exclude it . . .

Mr McKenna: Yes, I exclude it.

May was committed for trial. When he appeared at the Old Bailey on 1st May 1946 he was described by one newspaper reporter as being 'a slightly built man, he has thinning brown hair and small moustache'. When charged under the Official Secrets Act of communicating to a person unknown information which was calculated to be or might be useful to an enemy, he responded after a pause with, 'Guilty'.

The 34-year-old man before the court cut neither a dashing nor a dangerous figure. In his past his political, scientific and other intellectual pursuits as a young man could perhaps be described as unusual so far as the common run of men went, but seemed by no means extraordinary for a Cambridge student whose formative years coincided with the period of foreign interest and intervention in the Spanish Civil War and a general undergraduate sympathy with the left-wing cause.

After a distinguished academic career, which included a double first in Mathematics and Natural Sciences and finally a Ph.D., in 1936 he paid a short visit to Russia. In Britain he continued to move in leftish circles and became a member of the editorial board of *The Scientific Worker*, the official publication of the National Association of Scientific Workers.

On May's behalf, his counsel in court, Mr Gerald Gardiner, quickly made the admission that it was a Russian to whom May gave his secret information. The defence chose to rest its case on the plea that what May did was done for the best of all possible motives: the love of man. Gardiner explained May's actions, and returned to the subject which had caused the magistrate to intervene so sharply at the last hearing:

> It is not possible to do justice to him without considering what was going on at the time. In February 1945 the British Army was mostly in Holland and the Russians were in the course of their drive to Berlin. It was customary to refer to the Russians as 'gallant Allies', and they were doing their fair share in the war. . . .

At this the Attorney-General, Sir Hartley Shawcross, who was prosecuting, sprang to his feet to say:

> There is no suggestion that the Russians are enemies or potential enemies. The court has already decided this offence consists in communicating information to unauthorised persons.

Shawcross refuted May's statement that the information he had handed over had been, 'mostly of a character which has since been published or is about to be published', and remarked that by no means all of it had been made public.

Gardiner expanded on May's theme that the prisoner had embarked on this affair, which had proved so painful to him, because it was a contribution he could make for the safety of mankind. Said the defence counsel:

> Doctors take the view that if they discover something of benefit to mankind they are under obligation to see it is used for mankind and not kept for any country or people. There are scientists who take substantially the same view and Dr May held that view strongly.

There was an important flaw in this analogy, which Gardiner, like May, had failed to take into account. The cure for a disease is almost invariably put to the good use for which it was purposefully devised by doctors; the first use for atomic energy which physical scientists, May among them, had intended, was that of a devastating weapon of war. Had May any right to expect sympathetic understanding when he so self-righteously proclaimed that he had set himself up to be the judge of whether any other State should be helped to the ownership of that weapon?

Gardiner concluded and May was given the opportunity to speak, but refused it. Mr Justice Oliver then pronounced his sentence:

> Alan Nunn May, I have listened with some slight surprise to some of the things which your learned Counsel has said he is entitled to put before me: the picture of you as a man of honour who had done only what you believed to be right. I do not take that view of you at all. How any man in your position could have had the crass conceit, let alone the wickedness, to arrogate to himself the decision of a matter of this sort, when you yourself had given your written undertaking not to do it, and knew it was one of the country's most precious secrets, when you yourself had drawn and were drawing pay for years to keep your own bargain with your

country – that you could have done this is a dreadful thing. I think you acted not as an honourable but as a dishonourable man. I think you acted with degradation. Whether money was the object of what you did, in fact you did get money for what you did. It is a very bad case indeed. The sentence upon you is one of ten years penal servitude.

Six and a half years later, after having been a model inmate and having earned himself full remission, Alan Nunn May, as Prisoner 6499, was released from Wakefield Gaol. A few days before the release was due, a former fellow prisoner earned himself a few pounds by selling his reminiscences of his days at Wakefield with 'the atom scientist'. The article had little to add to the evidence given in court except to state that May maintained throughout his stay in 'peter' that he never took a penny for his spying activities: 'Why should I? My father and mother both have money.' But not all the prisoners believed this story, and one tale which did the rounds of the prison was that May received £50,000 for the information he had passed on.

Looking forward to his release from prison he was reported to have said, 'Any good scientist can get money to finance a laboratory. But this time I shall know better. I am going to work and work just for the pleasure of finding out for myself.'

After his release he travelled to the home of his brother in Buckinghamshire and from there issued a statement, typed by his sister-in-law, which read:

I do not wish to discuss the details of the action which led to my imprisonment. I myself think that I acted rightly and I believe many others think so too.

To those who think otherwise I would like to point out that I have suffered the punishment which was inflicted on me by the law and I hope I shall now be entitled to at least the consideration normally granted to released convicts: an opportunity to restart life.

There is just one of many erroneous statements of fact which have been made about me which I should like to correct now. I was not convicted of treason, nor was this word used by the prosecution or judge at my trial and I certainly had no treasonable intentions. I was wholeheartedly concerned with securing victory over Nazi Germany and Japan and with the furtherance of the development of the peaceful uses of atomic energy in this country.

My object now is to obtain as soon as possible an opportunity of doing useful scientific work in which I can be of some service to this country and to my fellow men.

Now that my imprisonment is over I can only wish for the same consideration and fair treatment which I received throughout the long period of my sentence from the prison officials and my fellow prisoners.

For a time May worked in a commercial laboratory in Cambridge before taking up a professorship in physics at the University of Ghana in 1962. Both in Britain and in Africa he has maintained the silence, which his release statement said he intended to maintain, on the motives for the actions which turned scientist into convict.

The roots of the conscience which fed these motives are not easy to uncover from beneath the interpretations of prosecuting counsel, defence counsel, incomplete self-confession and judgment. They are without doubt twisted and confused.

In the signed statement he made after his arrest, and subsequently, Nunn May insisted that the humanitarian motive was all, and that the profit motive was in no way present at the time that he was handing over information to the Soviet Embassy staff in Ottawa; he maintained that the cash he had accepted in return had been against his will, In his defence his counsel had insisted that May had nothing to gain and everything to lose. However, as the Canadian Government Royal Commission reported, cipher telegram number 244 of 1945 from Grant (Colonel Zabotin) to The Director in Moscow, referring to arrangements made with May to contact another agent on a London street, includes the sentence, 'We handed over 500 dollars to him'. Also, in one of the notebooks carried out of the Soviet Embassy by Gouzenko is the handwritten sentence, '200 dollars Alek and 2 bottles of whiskey handed over 12.4.45'. It is signed by *Baxter* (Lieutenant Angelov) and is in an account of the meeting in which Nunn May handed over a sample of U_{235}. Nunn May's confession, however, admits having received 'some dollars (I forget how many) *in* a bottle of whisky'.

Whereas 700 dollars and a couple of bottles of whisky – assuming that this is the whole of the reward the physicist had for his information – is hardly a fortune and an alcoholic feast, it is far from being merely a token. To May, an Englishman working in wartime Canada, 700 dollars could represent a considerable increase in standard of living. If he was bent only on registering a bond of dependence with a new master, then he could have

231

achieved this small symbolism at less cost to the Soviet Embassy, and with less gain for himself.

Though his punishment was set at 10 years' imprisonment by Mr Justice Oliver, there was no way of assessing in real terms the damage he had done to the relations between Russia and the Western atomic powers. The information he passed to Russians was, at the time of his actions, a precious secret owned by America and Britain. Russia pooh-poohed the value of this information in a statement issued in Canada in February 1946 which claimed that the technical data it contained was outdated by advances already made in the U.S.S.R., and which pointed out that a great deal of knowledge had now been given freely by the publication in the United States in August 1945 of 'the well-known brochure of the American, J. D. Smyth [*sic*], *Atomic Energy*'. Some of the information was, nonetheless, passed to Russia before the publication of Henry Smyth's report and May's confessed purpose in passing it was to speed the dissemination of information on atomic energy for the safety of mankind. At the time of his espionage there was only one immediate use to which atomic energy could be put: the manufacture of an atomic bomb. That the safety of mankind would be improved by Russia having such a bomb more quickly as a result of receiving this information is the only interpretation that can be given to May's motive in this case.

But the information which May handed over was not limited only to details of progress towards utilisable nuclear energy. He had also been asked for, and had given, what little news he knew of some American electronically controlled anti-aircraft shells: conventional weapons of use only in wartime, and a knowledge of which could scarcely benefit mankind economically, or the world politically.

May, in taking up his post in nuclear research, had signed the Official Secrets Act; indisputably he violated the trust to which this action had bound him. To say, as he did on his release, that he committed no treachery and had not been convicted of such a dishonourable act might salve his own conscience, but does not alter the fact.

His misfortune was to be a man of his time: a scientist who reached his maturity among physicists who as a group were reaching a new kind of maturity: men who were accepting fearsome responsibilities in wartime because their contribution in an act of creation had made them peculiarly informed of the likely nature of the consequences of their creativity. May had accepted that he must bear his new responsibility, but he had interpreted the manner in which he should act, as a result, differently from most of his colleagues, and wrongly.

May believed that the special position in which being a scientist had placed him meant that he could take decisions and actions which would transcend the barriers governing the behaviour of ordinary mortals. He claimed that his work as a physicist was for the benefit of all mankind and that the behaviour which put him in prison was likewise for the benefit of all mankind. He was as bitterly hurt by the interpretation placed on this behaviour as by the punishment he was made to suffer as a result of it.

In the long run, and in the light shed by 20 years of atomic diplomacy, the total harm done by May's treachery is negligible. There has been no war in which nuclear weapons have been used and nothing that May did or said could possibly influence the outcome of such a war, even were it to develop in the near future. At the time when it was discovered that he had passed information there were some fears expressed that the knowledge would speed Russia on her path to the atomic bomb, but this was doubtful even then.

Nor was May alone in believing that it was of paramount importance that atomic secrets be shared with Russia, our great ally at that time. As again voiced at the trial, this was a widely held view among scientists. But if May was aware of the efforts being made in 1945 in both the United States and Britain by such men as Niels Bohr and Robert Oppenheimer to bring Russia into the confidence of physicists and politicians alike, then he never spoke of it. Instead, he interpreted his singular position as one which entitled him to move outside the conventional channels other scientists were using and to take unilateral action. To his surprise and bitter regret he found the scientist was punished as an ordinary mortal.

May still lives and works in Ghana as a professor of physics, still apparently holding to his statement that, 'I myself think I acted rightly'. Gouzenko, even now still fearing reprisals for *his* actions, lives secretly near Toronto; in addition to his son, now in his middle twenties, he has a daughter three years younger than her brother. Occasionally Gouzenko appears at meetings to discuss what he hated most in life in a communist state; he wears a mask over his head to keep safe his identity.

On 5th April 1950 the Lord Chancellor, Viscount Jowitt, rose in the House of Lords. In the previous week, he reminded his peers, he had said that there was no truth in certain statements made in the press arising from a recent spy case in which it was suggested that there was a link between that case and the case of Alan Nunn May. Now he wished to correct himself, to apologise to their Lordships, and to the press. The fact was that in a notebook belonging to a man who was one of those examined by the Canadian Royal Commission after the Gouzenko affair, there had appeared, among a long list of names and addresses, the name of the central figure in the recent case: that of Klaus Fuchs. 'Subsequent events,' said the Lord Chancellor, 'have, of course, attached a significance to that name which it did not then bear.'

The owner of the notebook was Israel Halperin, a professor of mathematics at Queen's University, Ontario, who was born of Russian parents. Under the cover-name 'Bacon', Halperin had been mentioned by the Gouzenko papers as a reluctant member of one of the Russian spy-rings operating in Canada. Though eleven Canadian citizens were eventually convicted of espionage, Halperin was acquitted and continued to live an honourable life in Canada with such useful effect that in 1967 he was awarded one of four medals of the Royal Society of Canada for outstanding achievement in the physical sciences, along with a cash award of 1,000 dollars.

Viscount Jowitt's apology to the House in 1950 confessing ignorance of the May–Halperin–Fuchs link was being given four months after the arrest of Fuchs and over three years after Fuchs' name had been found in Halperin's list. The blunt implication was that British and Canadian security authorities had been far from speedy in following up obvious leads which might reveal breaches in the security-wall of member nations of the Commonwealth.

The Lord Chancellor's speech exposed yet again the disadvantages which a modern industrial society seemingly must suffer by, of necessity, placing such great trust and responsibility in the hands of an élite group of scientists. Here a small clique of them, mathematicians and physicists, had apparently been able to put the safety of the State in jeopardy. However, the links between the scientists, in this case at least, were at best tenuous and Halperin had perhaps done no more than send Fuchs periodicals and magazines by mail. But in 1950 the atmosphere of suspicion towards the group of men who had shown themselves to be so indispensable to their nation during war was approaching its most intense. The case of Klaus Fuchs was a bitter pill to swallow both for Britain, his country of adoption, and for America, whose hospitality he had received as a scientist.

When he judged the right time had come, Fuchs made the facts of his life from childhood to arrest freely available with a surprising fullness and with an obsessive attention to the minutiae of the workings of his mind in order to explain his actions. He was born in a village between Frankfurt and Mannheim in 1911, the son of Emil Fuchs, a man of deep religious conviction who became first a Lutheran pastor and who later, when Klaus was a young boy, joined the Society of Friends. When Klaus entered university he quickly keyed himself into the protesting undergraduate politics of the late 'twenties; he joined the Social Democratic Party, with which his father was now associated, and became a member of the Reichsbanner: a militant political organisation with many young social democrat members.

As he became more deeply involved in student political organisations and, as the fascist–communist conflict developed, he found his sympathies drifting further to the left. As chairman of an organisation of mixed social democrats and communists, he quickly got in conflict with the Nazi party, and before long had himself made a communist party member.

Expelled by the Democrats, he was now in danger of being at the receiving end of violence from the Brown Shirts. With the burning of the Reichstag and the release of the great wave of Nazi brutality and murder, Fuchs knew that the period of his overt support for militant communism was at an end; on the day that he read of this violent affair in a newspaper, at once he removed the visible sign of his allegiance: a hammer and sickle which he wore in the lapel of his jacket. For Fuchs this act of hiding a painted piece of metal was a symbol of deep significance which he took some care to recall 17 years later; for during the whole of those years he was to keep hidden from all of those who believed themselves to be in his confidence the true nature of his political commitment.

He went into hiding, and the communist party made arrangements to send him abroad. After a short spell in France he arrived in England and went to Bristol University to complete his interrupted studies for his Ph.D. in mathematics and physics. Fuchs later learnt that during this period his father, the staunch Christian and socialist, was going through a period of intense suffering at the hands of the Gestapo. When the Nazis came to power the old cleric was imprisoned for a time. On his release, in spite of being under Gestapo surveillance, with his remaining sons he operated a car hire business as a cover to help escaping refugees. But before long these sons, like Klaus before them, had to flee the country. Klaus's sister helped her husband also to escape; but her mind suffered during this time of strain: she committed suicide by throwing herself from a train.

One attempt was made to draw the attention of British authorities to Fuchs's political allegiance during his early stay in England. Whilst he was in Bristol the German Consul unofficially informed the Chief Constable that Fuchs was a communist. But already during these early years of the Hitler régime, deliberate attempts had been made on behalf of other German Consuls in Britain to prejudice the position of German refugees by making unfounded allegations against them. Fuchs was not then, and never did become, a member of the British communist party, so that it is scarcely surprising that no particular interest was shown in the news.

Having had his doctoral thesis accepted at Bristol, he took up a research scholarship at Edinburgh under another émigré, the distinguished physicist, Max Born. And when, with the beginning of war, Britain interned most of her aliens, Fuchs was sent off to Canada to a camp near Quebec, since described by other inmates as having a strong communist core.

In 1941 he was given permission to return to Britain where soon he joined Rudolf Peierls at Birmingham University; he thus became one of a long line of highly intelligent refugee physicists to find employment on the atomic bomb project when other highly secret activities were excluding alien physicists. Like the rest brought into that project he signed the Official Secrets Act, and like most became a naturalised Briton.

In November 1943, as one of Peierls's assistants, Fuchs became one of 19 British scientists to travel to the United States to join the Manhattan Project. At Columbia University in New York City he worked on theoretical aspects of the gaseous diffusion plant, then later joined the British contingent at Los Alamos where in the course of his theoretical contributions he became privy to many of the details of the uranium and plutonium weapons. He was one of the senior scientists who was taken to witness the Alamogordo explosion.

Several months later, the war over and the taut purposeful atmosphere of the atomic village relaxed, a dinner party was held in Los Alamos which Fuchs and some of his colleagues and their wives attended. These parties in the isolated prefabricated group of buildings on the mesa were not uncommon, but this one had a special interest: it was held on the day of the arrest of Alan Nunn May. The physicist, Edward Teller, was also present at the dinner and later recalled that he noticed that Fuchs appeared to have been deeply affected by the news of the arrest: so much so that Teller remarked on it to his wife on their journey home. Fuchs, well known for his lack of effusiveness, had earned himself the name 'Penny-in-the-slot Fuchs' from Mrs Peierls who felt that she always had to put a sentence into the quiet bachelor to get a sentence out of him. On this occasion, with some fellow physicists doubting May's guilt and others arguing that spying by some of their colleagues was an inevitability, Fuchs had appeared to Teller to be unusually affected by May's situation.

Fuchs returned to England in June 1946 and became the head of the theoretical physics division of the Atomic Energy Establishment at Harwell. He was holding that post when a telephone call on 2nd February 1950 took him for the second time in four days to the office at Shell-Mex House, London, of Michael Perrin, the Deputy Controller of Atomic Energy (Technical Policy), at the Ministry of Supply. On this occasion he was arrested on charges made under the Official Secrets Act.

As with Alan Nunn May five years before him, Fuchs faced two hearings at Bow Street Court, and then a short, sharp trial at what was described during the proceedings as the 'Senior Assize of the Empire' – the Old Bailey. Again this case of the fractured loyalty of a physicist roused a great deal of public interest, for its implications seemed of grave world importance; 80 newspapers were represented at the trial, and a seat provided for Tass, the Russian news agency. Both hearings and trial excited the British press to extract from the histrionic headlines, such as, 'Atomic Plant Jekyll and Hyde "Broke Oath to Britain" ' , 'Fanatic on Payroll of Foreign Power', and 'The Mind and Heart of a Betrayer'. And in the days following Fuchs's arraignment reporters sought out his co-workers, landladies, inn-keepers and the other few acquaintances they could find to try to gain some insight into the character of this softly spoken, bespectacled immigrant scientist whom they had seen in the dock.

A diffuse image of the man emerged. He was reticent and difficult to draw into conversation. He spoke little about politics beyond making clear his hate of the Naziism from which he had suffered. In general he shunned the deep human relationships which might have led to some earlier suspicions of his hidden convictions. Those who had known him had liked him, and had admired him for his scholarly devotion to his physics and to Harwell, the mushrooming laboratory on the edge of the Berkshire Downs. But there was to be nothing so revelatory to appear concerning Klaus Fuchs as that which he himself chose to reveal from chinks in his own protective shell, and which were the confessions he made to security officers before his arrest.

The charges he had to face were made under the Official Secrets Act of 1911. He was accused, for a purpose prejudicial to the safety or interests of the State, of having communicated to a person unknown, information relating to atomic research which was calculated to be, or might be, useful to an enemy. Four occasions were cited.

Fuchs had made both oral and written confessions and had given to the security officers, who investigated him, full details of the information he had passed to Russian agents from 1942 to 1949 when he had been successively a member of Tube Alloys in Birmingham, one of the British atomic group of scientists in the United States, and finally in employment at Harwell.

The only reward he received from his Russian contacts for this work was £100 soon after he returned to England in 1946, but this, he claimed, had been merely a token payment. He had discussed with a friend, who knew Alan Nunn May, May's acceptance of money, and as a result of this conversation he decided to follow suit and let these few pounds act as a symbolic payment which would signify his subservience to the cause.

'This is a case,' said Sir Hartley Shawcross for the prosecution, at Fuchs's trial[2] on 1st March 1950, 'of the utmost gravity, I suppose as serious as any that has ever been prosecuted under this statute.' He went on, 'That the information communicated was likely to be of the utmost value to an enemy is unhappily a matter which admits of no doubt.' But Shawcross was quick to point out that the country to whom the information was transmitted, in this case the Soviet Union, might never become an enemy and added, 'Our relations with that country leave much to be desired, but they are not those of enmity. Everyone hopes that eventually wiser counsels will prevail and that that country will live in amity and agreement with the rest of the world.'

The basis on which the Attorney General's case for the prosecution comfortably rested was the long statement which Fuchs had made; in this the covert communist's motives, as he saw them, were intertwined with the facts as he saw them. The man to whom Fuchs had dictated this statement was William Skardon of the Security Service, who in the past weeks had been responsible for drawing out of Fuchs all the thoughts which, because of his choice of an ideologically lonely servitude of the cause, had been bottled up in the past 17 years.

In a long interview with Fuchs in the previous December Skardon had brought the topic of conversation round to the oath of allegiance which Fuchs had taken on his naturalisation as a British subject. Fuchs had said that he regarded that oath as a serious matter, but claimed the freedom to act in accordance with his conscience, should circumstances arise in Britain comparable with those which existed in Germany in 1932 and 1933, when he would feel free to act on a loyalty which he owned to humanity generally.

After Skardon heard the disturbing news from this trusted Harwell physicist – that he of his own accord might choose to judge how best to serve humanity with the knowledge which as a scientist he had at his disposal – several weeks of waiting passed; then the confession flowed. It all began, said Fuchs, of his own initiative. At first he tried to restrict the information he divulged to telling Russia that an atomic bomb was being built and to handing over only that which he believed to be the product of his own brain. Perhaps he realised the impossibility of there being such a desirable ideal quantum of creation of the human mind, or perhaps he recognised the impossibility of his living with only partial commitment to his cause. Whatever the reason might have been, he soon began to give away freely other men's flowers, as he told in his statement, parts of which were read out in court:

> At this time I had complete confidence in Russian policy and I had no hesitation in giving all the information I had. I believed the Western Allies deliberately allowed Germany and Russia to fight each other to the death.

To Klaus Fuchs's acquaintances the solace of his life had been his work: his natural philosophy, his physics. But to Fuchs, that panacea which could stabilise the confusion in his inner self was his political philosophy: his communism.

> The idea which gripped me most was the belief that in the past man has been unable to understand his own history and the forces which lead to the further development of human society; that now, for the first time, man understands the historical forces and he is able to control them, and that therefore for the first time he will be really free. I carried this idea over into the personal sphere and believed that I could understand myself and that I could make myself into what I should be.

Fuchs even applied his communism in his confession in a flawed attempt at a self-analysis of his behaviour:

> . . . I began naturally to form bonds of personal friendship and I had to conceal from them my own thoughts. I used my Marxian philosophy to conceal my thoughts in two separate compartments. One side was the man I wanted to be. I could be free and easy and happy with other people without fear of disclosing myself because I knew the other compartment would step in if I reached the danger point. It appeared to me at the time I had become a free man because I succeeded in the other compartment in establishing myself completely independent of the surrounding forces of society. Looking back on it now the best way is to call it a controlled schizophrenia.

This consciously nurtured dual personality was less flatteringly referred to by Mr Christmas Humphreys for the prosecution, at one of Fuchs's appearances in court, as 'a classical example of the immortal duality in English literature, Jekyll and Hyde'. Just as Fuchs realised that the creative processes of his own mind could not be independent of those of other men, so too he now began to realise that in no sense could man be an island, and that the rules he had invented which would make him an independent unit of his chosen society were not now, and had never been, valid.

When the war ended and Fuchs returned to Harwell, the doubts about his own self-sufficiency became more intense and found expression in an attempt to sift only certain information through to his Russian contacts. Fuchs had difficulty in describing exactly how he carried out this process of sifting because, he said, 'it was a process which went up and down with my inner struggles'. It was his doubts about Russian policy which gave rise to these qualms.

> . . . eventually I came to the point when I knew I disapproved of many actions of the Russians. I still believed Russia would build a new world and that I would take part in it. During this time I was not sure I could give all the information I had. However, it became more and more evident that the time when Russia would spread influence all over Europe was far away. I had to decide whether I could continue to hand over information without being sure I was doing right. I decided I could not do so.

It was the news he received whilst he was working at Harwell, that his father was planning to go to live in the Eastern Zone of Germany, which brought Fuchs to his inner crisis. He loved his father and admired his sense of right and wrong. He believed that his father thought that communism was trying to build a new world, but having gone to that new world his father would not shrink from criticising the unnecessary sacrifice of human freedom used to build it. Dissatisfied with having to live a life of duplicity which the 'schizophrenic' compartments of his mind forced on him, Fuchs could not reach a decision whether or not he too should move to the new world.

> I suppose I did not have the courage to fight it out for myself, and therefore took it out of my hands by informing the authorities that my father was going to the Eastern Zone. A few months passed and I became more and more convinced that I had to leave Harwell.

Although Fuchs was ready now to believe and to confess that the worst of his actions had been to give away the principle of the design of the plutonium bomb whilst he was working as a trusted theoretical physicist at Los Alamos, what clearly had become the strangest of his contemporary obsessions was the belief that his revelations would totally undermine the stature of the atomic energy establishment at Harwell.

> I was then confronted with the fact that there was evidence that I had given away information in New York. I at first denied the allegations made against me. I decided I would have to leave Harwell, but it became clear that in leaving Harwell in these circumstances I would deal a great blow to Harwell and all the work I had loved and also leave suspicions against friends whom I had loved and people who thought I was their friend.

He was desperately concerned that the friends whose confidences he had betrayed should recognise the dilemma which had caused him to build the two compartments of his mind and, illogically, that they should judge him now by that entity on which was built his personal relation of love and trust. His statement ended:

> Before I joined the project most of the English people with whom I made contact were left-wing and affected by a similar philosophy. Since coming to Harwell I have met English people of all kinds, and I have come to see in many of them a deep-rooted firmness which enables them to lead a decent way of life. I do not know where this springs from and I don't think they do, but it is there.

To conclude his case for the prosecution at the Old Bailey, Sir Hartley Shawcross chose to emphasise that Fuchs's testimony had been obtained from him before his arrest when he was still a working Harwell scientist. Said Shawcross:

> I have had occasion before the United Nations to observe more than once that the courts of this country would not act upon so-called sinister confessions extracted in one way or another after a long period of secret incarceration and incommunicado, and by methods one knows not of, which have become a characteristic of proceedings in certain foreign countries.
> It should perhaps be said that this man's confession was made while he was still a free man, able to come and go as he chose and to consult with his friends and take the advice of his lawyers.

In Fuchs's defence, his counsel, Mr Derek Curtis Bennet K.C., pleaded that both the nature of the times and the nature of the man should be taken into consideration; he submitted that Fuchs ought not to be blamed too much if, during the fighting in the war, he gave information and then when Russia ceased to be our ally he went on giving information. Fuchs, he said, did not come to Britain to build atom bombs, and added, 'if the war had not come he might have been more a candidate for a Nobel peace prize or a membership of the Royal Society'.

Fuchs was convicted, and when asked if he had anything to say before sentence said in a low voice: 'My Lord, I have committed certain crimes, for which I am charged, and I expect sentence. I have also committed some other crimes, which are not crimes in the eyes of the law; they are crimes . . .'

At this point Fuchs's voice became so indistinct that reporters in the court were unable to hear him say that these other crimes were crimes against his friends. The reason he had put the facts before the court, he went on, was in order to atone for these other crimes.

Lord Goddard, the Lord Chief Justice who it was noted by the *Manchester Guardian* called the prisoner 'Foos', summed up what he believed to be the four matters which constituted the gravest aspects of Fuchs's crimes. First he had imperilled the right of asylum customarily given by Britain, so that all future refugees would be looked on with suspicion. Second, by his betrayals, he had caused suspicion to fall on the co-workers who had trusted him. Third he had imperilled relations between Britain and the United States, and fourth he had done irreparable and incalculable harm to both these countries. Lord Goddard ended by accepting that what Fuchs did was:

> . . . merely for the purpose of furthering your political creed, for I am willing to assume that you have not done it for gain.
>
> Your statement, which has been read, shows to me the depth of self-deception into which people like yourself can fall. Your crime to me is only thinly differentiated from high treason. In this country we observe rigidly the rule of law, and as technically it is not high treason, so you are not tried for that offence.
>
> I have now to assess the penalty which it is right I should impose. It is not so much for punishment that I impose it, for punishment can mean nothing to a man of your mentality.
>
> My duty is to safeguard this country; and how can I be sure that a man, whose mentality is shown in that statement you have made, may not, at any other minute, allow some curious working of your mind to lead you further to betray secrets of the greatest possible value and importance to this land?
>
> The maximum sentence which Parliament has ordained for this crime is fourteen years, and that is the sentence I pass upon you.

On 2nd March 1950, the day that it carried the news of the conviction and sentence of Klaus Fuchs, in a few relatively inconspicuous paragraphs, *The Times* reported from Washington the news that conversations between the United States, Canada and Britain on cooperation in the development of atomic energy had been suspended, and that this could be regarded as a direct consequence of the Fuchs case. This, and several other effects of the physicist's treachery, during that period when the fear of a war of atomic weapons was so strongly influencing the international political scene, was pushed into the sidelines of public concern. The immediate worry was how far and how fast Fuchs had managed to push Russian work on the atomic bomb.

In its pamphlet on Soviet Atomic Espionage, published in 1951, the U.S. Joint Committee on Atomic Energy noted that both General Groves and Robert Oppenheimer estimated that Fuchs might have set ahead the Soviet project by one year. It said:

> It is hardly an exaggeration to say that Fuchs alone has influenced the safety of more people and accomplished greater damage than any other spy, not only in the history of the United States, but in the history of nations.

Would such a sweeping assessment of the consequences of Fuchs's actions have been made had he faced trial at different times between when he was first invited to join the atomic project and the present day? For example, would the safety of the peoples of the Western nations have been considered to have been in such precipitous danger if his spying had been judged at any time during the 15 months when that ally of these peoples, the Red Army, was sustaining 5,000,000 casualties fighting Nazi Germany? Or, would the help he gave Russian scientists have been valued at such a high price during that period when the first Sputnik was reaching its apogee and demonstrating that Russian technology had progressed in some spheres further and faster than its Western counterpart, and without the benefits of cooperation with Western scientists which had been considered by some to be *sine qua non*?

It might well have come to pass that the greatest harm done by the Fuchs case would have been to cause panic political or even military action as a result of an over-assessment of the value of his disclosures and the likely consequences of these. Fuchs's treachery was uncovered at the time when the United States government was weltering in indecision whether or not to arm itself with a weapon to dwarf the atomic bomb. If the President had any doubt in his mind about the decision he had made to direct that work should go ahead with the hydrogen bomb, that doubt must surely have been removed by Fuchs's confessions.

As the years passed and as the clouds of danger which had been precipitated over the heads of the Western peoples were now beginning to disperse, so it became clear that the tangible harm done by Fuchs in the long run lay in the apparently minor side-effects of his actions.

Not only did he litter the approaches to cooperation between nations in atomic energy matters, he also unwittingly poured cold water on the efforts many of his erstwhile colleagues were making to secure international scientific cooperation in wider non-nuclear fields – so making more remote that desirable spread of scientific knowledge for humanity's good which he on his own initiative had tried to effect. At the same time, as Lord Goddard had pointed out at his trial, he brought under suspicion the motives of many leading scientists who had been involved in wartime work, in particular those who were émigrés to Britain and America, and who were amongst those who were most sensitive to the undesirability of national scientific barriers.

But perhaps one of the most damaging consequences of what was heard at Fuchs's trial was the attitude it caused to be fixed on the scientific community as a whole. Fuchs (and also Alan Nunn May) had helped establish a myth that scientists were a race apart from that of ordinary men, that their culture was not a part of the common culture, and that they therefore believed that they could act, and indeed had acted, with arrogance and conceit according to a set of rules which applied to them alone. This attitude was to affect the careers of several leading American scientists in the few years following the trial of Klaus Fuchs.

The public at large had heard with some surprise Fuchs's obsession that his beloved laboratory, Harwell, would suffer when it was revealed that he was passing on its secrets. Were there then other physicists who were equally concerned that the integrity or prosperity of a scientific institution should be more sacrosanct than the safety of the nation which housed that institution? Was Fuchs's peculiar logic common to others of his scientific kind? Were scientists a separate breed who should be treated accordingly? Fuchs's defence counsel, Mr Derek Curtis Bennett K.C., at his trial had attempted to answer this last question in order to explain why Fuchs had continued to pass information to Russia when the war was over and when it seemed that the country might turn from ally to enemy. He had said:

> A scientist is in this position: he is taught, or teaches himself, or learns, that A plus B equals C. If he is told tomorrow that it is A minus B that equals C, he does not believe it. But your sensible citizen or politician, moving in the affairs of the world, told that, would agree with both. He has to. But the change of political alignments is not the business of scientists, for scientists are not always politically wise. Their minds move along straight lines, without the flexibility that some others have. . . .

To illustrate the confusion in his scientific client's mind Curtis-Bennett had chosen what he believed to be an appropriate scientific analogy. In order to make it he had had to make the assumption that scientists are of a different quality from ordinary citizens and that therefore they should be judged as scientists and not as ordinary sensible citizens. But such an assumption is palpably untrue and dangerous. It is dangerous because it reinforces the conviction of men such as Fuchs that they are free, under certain circumstances, to behave with a different moral standard from that of their fellow citizens and, if necessary, to put in jeopardy the safety of the nation to which they have sworn allegiance.

In his closing remarks Mr Curtis-Bennett went on to say of Fuchs's repentance: 'It is perhaps a consolation to know that Dr Fuchs' attitude has been changed. And it has been changed as a result of his association with British people and chiefly British scientists whose humanity and decency have conquered him.'

Was one to assume, then, that British scientists who exerted their beneficial influence on Fuchs were more fully integrated into the common culture of humanity than were those of other nations?

Many of Fuchs's colleagues on the Manhattan Project, both British and American, made clear their condemnation of his treachery. Karl Cohen of Columbia University, in a letter to the Joint Committee on Atomic Energy, wrote to say that although the personal integrity of scientists which Fuchs had betrayed was only of minor importance, 'nevertheless it was a blow which all scientists bitterly resent'. The petitions from scientists for a remission of sentence which followed Alan Nunn May's conviction were not repeated in the case of Fuchs.

On 12th February 1951 the Home Secretary, Mr Ede, made an order depriving Dr Klaus Emil Julius Fuchs of his British citizenship 'on grounds of disloyalty'. In 1959, after serving nine years of his sentence, Fuchs was released from gaol and immediately boarded an aeroplane for Eastern Germany, leaving behind for good the country in which he had found so many people with the 'deep-rooted firmness which enables them to lead a decent way of life'. Within weeks he had taken up an appointment at the Central Institute for Nuclear Research at Rossendorf. As deputy director of this institute he gave an interview in 1965 in which he stated his belief that West Germany was developing nuclear power for military purposes. In the same interview he recalled his pleasant recollections of Harwell, but tempered that recollection with the memory of the distaste he felt when he realised that the primary purpose of that establishment was not, as he had been led to believe, a purely civilian and peaceful one, but was to produce an atomic bomb.

During the late summer of the year of Fuchs's exposure, another Harwell physicist was preparing himself for an action which, because of the manner in which he was to perform it, was to make the British and American public fear that yet another chink had appeared in the sanctuary wall of its post-war existence. This public was not to know what he was about until 21st October 1950 when they learnt of him from their newspapers for the first time, and learnt that his colleagues described him as 'one of the top people – a really clever scientist'. He was a lively Italian Jew named Bruno Pontecorvo who had been working at the Atomic Energy Establishment for 18 months.

Pontecorvo had taken his Ph. D. at Rome University and there had worked with a flourishing group of physicists, the pride of Italian natural science, which included Enrico Fermi and Emilio Segrè, both of whom were destined to become key-members of the Manhattan project. In 1935 this group published a paper, in the *Proceedings of the Royal Society* to which Fermi, Segrè, Pontecorvo and three other physicists' names were attached, called

'Artificial Radioactivity Produced by Neutron Bombardment'. As a result of this work the group successfully filed a patent, and after the war was awarded 300,000 dollars' compensation from the United States' government which had utilised it, infringing the patent. But in 1953 when the group's claim was assessed, and Pontecorvo's share of the award was set at one-sixteenth, he was not around to hear of his good fortune. His share was deposited in the United States Treasury.

Being a Jew he had, like others in the group, fled from Italy and Mussolini's brand of fascism in 1936. Eventually he was taken on as a member of the British Atomic Energy Project in wartime Canada, and, like Alan Nunn May, visited the Chicago Metallurgical Laboratory to discuss the progress of work there. He joined Harwell in 1949 after having taken British citizenship and appeared to have no regrets about the country of his choice. He had been offered and had accepted a professorship at Liverpool University and a distinguished British academic career apparently now stretched comfortably before him when he took off in his car with his Swedish wife Mariana and their three children in July of 1950 for a continental holiday.

The Pontecorvos wandered idly southwards through the countries of Europe, camping, swimming, underwater fishing and, to all outward appearances, relaxing before making family visits in Italy. Postcards to friends carried the customary salutations announcing 'having a wonderful time' and 'a lot of fun' before the carefree holiday turned into a scarcely disguised flight. From Rome Pontecorvo took his family by aeroplane to Stockholm; in that city lived Mariana Pontecorvo's parents, but on this visit neither Bruno nor his wife made any attempt to contact them. There, on 2nd September, the family boarded a plane for Helsinki and on arrival transferred its luggage to a waiting car; at this point the trail, to Western eyes, petered out. There were several reports as to how the Pontecorvos left Finland. One was that the car drove them to Russian-leased territory only 12 miles outside the capital; another was that the departure of the Soviet ship *Belostov* from Helsinki was delayed before sailing for Leningrad until the Pontecorvos were aboard.

The news that Pontecorvo was behind the Iron Curtain was broken by all except communist newspapers in Rome on 20th October and the story was carried by British newspapers on the following day. Although most of the newspaper stories were speculative, there was never much doubt that the family had ended up in Russia, and Mr Strauss, the Minister of Supply, had to say as much in answer to worried questions in the House of Commons. But it was not until March 1955 that Pontecorvo himself chose to reveal his situation. In an article in *Pravda* and *Izvestia* he described how he had 'just emigrated' to the Soviet Union from Britain and how he was now working on the peaceful uses of atomic energy. In Britain, he said, he had been subjected to 'direct questioning and systematic blackmail on the part of the police authorities' and that this had convinced him that he could 'no longer preserve my dignity if I remained where I was'. A few days later, wearing the medal and ribbon of the Stalin Prize, he made his first public appearance at a press conference at the Academy of Sciences in Moscow where he disclosed that he had become a Soviet citizen in 1952. He was reported by the *Daily Worker* to have said there, 'I believe that many British scientists and physicists, men of honest conviction are allowing themselves to be blinded by the yellow press. Only thus could I explain their opinions about the Soviet Union.' The removal of his British citizenship was announced in May 1955.

For several years he worked at the Joint Institute of Nuclear Research at Dubna near Moscow, and in 1963 he was awarded the Lenin Prize of 7,500 roubles (about £3,000) for his work on the physics of the neutrino and the weak interactions between elementary particles. Tass, making the announcement, called him one of the outstanding physicists of our time. Thus, in a fashion, he was compensated for those moneys which still lie in the United States Treasury. In August 1967, Tass announced that Pontecorvo had been appointed head of a new school of space physics in Siberia.

The two questions concerning Bruno Pontecorvo which have never been satisfactorily answered are: why did he leave Britain, and why did he leave in the way he did? At the press conference in 1955 he gave his own version of the answer to the first question but baulked at a reporter's probing of the manner in which he got to Russia, which might have led to an answer to the second. But if, as he claimed, the reason for his flight was a result of questioning and blackmail by British security, why did he quit the country if he had nothing to hide? Had he passed information to Russia during or after the war? Had he had some espionage contact with Klaus Fuchs? If not and if he had a clear conscience, then, being the intelligent man he was, surely he could not have believed that Police State methods would be used to endanger his life or restrict his freedom.

If he had not been a spy, did he then have in his possession or in his brain some information to take with him which might be valuable to Russia? The general consensus of scientific opinion is that Pontecorvo did not have access to such valuable property as would dictate such a dramatic departure for the East; that his existing knowledge could not have made a radical difference to the Russian nuclear or thermonuclear weapons effort; and that his greatest virtue to the Soviets would be for his expertise as an experimental physicist – or for his propaganda value.

Is it possible that his work on atomic weapons' research in Britain so troubled his conscience that he wished to dissociate himself from this distasteful application of his scientific abilities? Statements he has made from Moscow appealing to Western scientists to take a stand against nuclear armament research suggest that this might well have been a strong motivating factor. But in that case, why did he choose the time to high-tail it out of the country when he had just accepted a university job? And again, why in the manner that he did? If his loyalty to Britain until the moment of his departure was beyond suspicion, why did he not wait until a more favourable time, when he could have 'just emigrated', as he called it, in a less stagy fashion?

Or is it possible that there was a more mundane reason for his absconding? During that European summer holiday, did some family affair reach an emotional climax which would make him want to build the rest of his life in a new context?

It is possible that Pontecorvo or his sponsors might one day wish to reveal the answers to some of these questions. Because they were so unanswerable at the time of his flight in 1950, an aura of mystery and wary, even fearful, speculation surrounded his outwardly clumsy journey from Harwell, through Rome and Helsinki, to Moscow. Once again, a nuclear physicist – to the layman an enigmatic but valuable piece of scientific real-estate – had seemingly put the society which fostered him in peril. The only difference between him and the others of his ilk was apparently that he had not merely let the cat out of the bag: he had taken cat and bag off with him.

The disclosures which within five years laid open the lives and the treachery of May and Fuchs and the abscondence of Pontecorvo did little good to the exposed surface of the British system of security, and there was to be yet more tarnish on that surface before many more years were to pass. For a period, Americans whose secrets, along with those of Britons, had been betrayed, let loose some harboured bitterness; General Leslie Groves, the military head of the Manhattan Project, in his customary forthright fashion, left no doubt that he believed that rank incompetence was at the root of these security woes. He has since declared his belief that the statement of reliability of the British scientists who worked under his command, given to him and confirmed at his insistence by a British official, was merely a sop, and that no investigation by the British of Fuchs's background was made at all.

America, however, could not sit smug for long. During a period of a few months in 1950, the year of Fuchs's conviction, nine Americans, all linked with his case, were arrested. All of them had some background of, interests in, or direct connections with science or technology, and whereas none of them could be described as being first-class scientists, or even aspirants to that rank, and some were technicians rather than technologists, they nevertheless added to the confused image which had now evolved of the scientist as a member of a queer breed and the owner of a vulnerable loyalty.

The first link from Fuchs was a small and particularly undistinguished one named Harry Gold: double-chinned and overweight who in his own unprepossessing turn of phrase was 'sloppy fat'. Gold, a chemical engineer who had been born of Russian parents in Berne in 1910, had been brought to the United States as a child. He had been identified as Fuchs's contact man in America after F.B.I. agents had been allowed to visit Fuchs in Wormwood Scrubs; within three days of their first interview enough information had been pieced together to bring about Gold's arrest in Philadelphia and for him to give what was described as 'a detailed account of his activities'.

Gold, according to his own evidence, had been engaged in small-time industrial spying for Russia since as early as 1935. He had progressed from handing over information about processes used in the sugar company where he worked, through the passing-on of commercial photographic secrets, to acting as the go-between for Russian agents and Klaus Fuchs. He and Fuchs had met several times in New York City and in Cambridge, Massachusetts, in 1944 and early 1945, and in Santa Fé, a quiet town near Los Alamos, in June 1945. As a result of some of these contacts Gold had been able to deliver packets of information from Fuchs to Anatoli Yakovlev of the Soviet consulate.

Nine months after Klaus Fuchs began to serve his sentence of 14 years, Harry Gold received the longest prison term which a Philadelphia court was capable of awarding for his offences: 30 years.

Three months later Gold appeared as a prosecution witness at another espionage trial in a New York court and gave evidence how on 3rd June 1945, after having collected a bunch of papers from Fuchs in Santa Fé, he took a bus to neighbouring Albuquerque where next day, obediently following the instructions of Yakovlev, he contacted a 23-year-old Los Alamos machine-shop army technical sergeant named David Greenglass.

According to the prosecution at this trial, in exchange for $500, Gold got from Greenglass certain sketches and material which described an experiment with a device known as a *high explosive lens mold*, which was used at Los Alamos to form explosives into critical shapes. He also handed over a list of names of workers who might be suitable candidates for espionage. The pivotal figure who, it was claimed, had been responsible for bringing Greenglass into contact with the Russian espionage net, was one Julius Rosenberg, whose wife, Ethel, was Greenglass's brother. The trial at which this involved information chain was exposed was that of Julius and Ethel Rosenberg.

Named as co-defendent in the trial with the Rosenbergs was Morton Sobell, who was accused of conspiracy to commit espionage and of acting as another link in the Rosenberg spy chain. Julius Rosenberg, who had graduated from New York's City College with a degree in electrical engineering, had been removed from his post as civilian inspector in the Army Signal Corps in 1945 as a result of allegations that he had been a communist party member. Sobell, also an electrical engineer, was a graduate of the same college, and had taken a master's degree at the

University of Michigan; at the trial evidence was given that he had worked for the Navy's Bureau of Ordnance, and that he too had been a member of the communist party.

The trial took place during that most fearsome of phases when, in the 'fifties, Americans were subjecting themselves to a punishing orgy of indiscriminatory anti-communist political activity. The events responsible for this phase were strongly emotive ones. Abroad, General MacArthur, as commander-in-chief of United Nations forces, far from having brought the war in Korea to a brief close, had had to face what he called a 'new war': a prolonged winter offensive by a Chinese army. During this time the news from Britain of Fuchs's admitted treachery on behalf of China's ally in communism, Russia, was fresh in mind. At home, the hounding activities of Senator Joseph McCarthy, attended by a credulous public, and the prospects that the armament range available in any future full-scale war might have its spectrum completed by the addition of a hydrogen bomb, together played on the nation's fear of the consequences to be expected from subversive left-wing activities.

The trial, which opened on 6th March 1951, began with the prosecution claiming that evidence would show that the loyalty of the Rosenbergs and Sobell was not to the United States, but to international communism. It closed with all three being found guilty by jury, and being awarded the most stringent punishment. Judge Irving Kaufman said in his summing up:

> The competitive advantage held by the United States in super-weapons has put a premium on the services of a new school of spies – the home-grown variety that places allegiance to a foreign power before loyalty to the United States. The punishment to be meted out in this case must therefore serve the maximum interest for the preservation of our society against these traitors in our midst. . . .
> . . . I believe your conduct in putting into the hands of the Russians the A-bomb years before our best scientists predicted Russia would perfect the bomb has already caused, in my opinion, the Communist aggression in Korea, with the resultant casualties exceeding fifty thousand and who knows but that millions more of innocent people may pay the price of your treason.

He sentenced the Rosenbergs to death and Sobell to the maximum prison term of 30 years. In spite of a vigorous campaign for clemency outside the United States, on 19th June 1953 the

Rosenbergs died in the electric chair in Sing Sing; both protested their innocence to the last. Sobell, released in January 1969 after serving 18 years in a federal penitentiary, also still claims to be innocent of the charges made against him. The co-conspirator, Greenglass, who along with Gold had confessed and acted as one of the key witnesses for the prosecution, told by Judge Kaufmann that his help was recognised 'in apprehending and bringing to justice the arch criminals in this nefarious scheme', was awarded 15 years in prison. He was released in 1960.

Shortly after the executions a French biochemist and later a Nobel Laureate, Jacques Monod of the Pasteur Institute, wrote to the American *Bulletin of the Atomic Scientists* and spoke of the profound emotion which, especially in France, the execution of the Rosenbergs had aroused. He emphasised the unanimity of the desire for mercy which had been expressed in his country among individuals and groups which had included both eminent scientists and writers. He attempted to review the worrying points in the case which had appeared most significant to French intellectuals: that the basis of the prosecution case had rested on the testimony of avowed spies such as Gold and Greenglass, and that a 'simple mechanic' such as Greenglass, under the direction of an equally untrained Julius Rosenberg, would not have been capable of assimilating and memorising decisive atomic secrets. He quoted a letter to President Eisenhower from the American Nobel prizewinner, Harold Urey, in which Urey had said, '. . . Greenglass is supposed to have revealed to the Russians the secrets of the atomic bomb. Though the information supposed to have been transmitted could have been important, a man of Greenglass's capacity is wholly incapable of transmitting the physics, chemistry, and mathematics of the atomic bomb to anyone.' Monod thus found difficulty, like many others before and since, in accepting Judge Kaufman's contention that the A-bomb had been given to the Russians years before they could reasonably have been expected to produce it for themselves. Monod went on,

> . . . the gravest, the most decisive point was the nature of the sentence itself. Even if the Rosenbergs actually performed the acts with which they were charged, we were shocked at a death sentence pronounced in time of peace, for actions committed, it is true, in time of war, but a war in which Russia was an ally, not an enemy, of the United States. As outsiders to both countries, we

French could not help comparing this sentence with the six years given Alan Nunn May, and the thirteen years given Klaus Fuchs in English courts, for acknowledged and capable atomic espionage that the Rosenbergs could not have undertaken. . . .

He had hoped that American intellectuals and men of science would act as a handful of French intellectuals had done in the Dreyfus case when speaking out against a technically correct decision of justice. He continued,

> American scientists and intellectuals, the execution of the Rosenbergs is a grave defeat for you, for us, and for the free world. We do not for a moment believe that this tragic outcome of what appeared to us a crucial test case, means that you were indifferent to it – but it does testify to your present weakness, in your own country. Not one of us would dare reproach you for this, as we do not feel we have any right to give lessons in civic courage when we ourselves have been unable to prevent so many miscarriages of justice in France, or under French sovereignty. What we want to tell you is that, in spite of this defeat, you must not be discouraged, you must not abandon hope, you must continue publicly to serve truth, objectivity, and justice. If you speak firmly and unanimously you will be heard by your countrymen, who are aware of the importance of science, and of your great contributions to American wealth, power, and prestige. . . .[3]

A letter from an American published in reply spoke of the pride the United States had in its country's strength in carrying out the laws of the land – 'we are awake to the evils of communism where the French intellectual is still soft to the problem – this may explain the tragedy of the last war as far as France was concerned'.

In 1965 a book by Walter and Miriam Schneier called *Invitation to an Inquest* cast doubt on the reliability of some of the evidence given at the Rosenberg–Sobell trial. As a consequence, Morton Sobell's lawyers attempted, during his sixteenth year of imprisonment, to have a new hearing authorised for him: it was their seventh unsuccessful attempt to that date. Attached to Sobell's attorneys' motion to reopen his case were affidavits from two wartime Los Alamos scientists, Dr Philip Morrison and Dr Henry Linschitz, now both highly respected academics.

The severity of the sentences on Sobell and on the Rosenbergs had been based on the nature of the information passed to Russia: information which, according to David Greenglass, the technician responsible for carrying it out of Los Alamos, gave 'a pretty good description of the atomic bomb'. At the trial some sketches done by Greenglass of high explosive lens molds and of the atom bomb were produced, but these were merely replicas of those he had passed on to Gold and Rosenberg and were drawn for the F.B.I. after his arrest in 1950.

In spite of the fact that the prosecution at the 1951 trial had listed a number of eminent atomic scientists, including Robert Oppenheimer, Harold Urey, and George Kistiakowski, among its potential witnesses, less well-known figures connected with Los Alamos were in the event produced to testify to the value of the information transmitted by Greenglass. It was the reliability of this scientific testimony to which Professors Linschitz and Morrison addressed their affidavits in the 1966 attempt to reopen the case of Morton Sobell.

Morrison attacked the trial evidence of the prosecution witness John A. Derry who, acting as a liaison officer for General Groves, had visited Los Alamos and had observed the development of the atomic bomb. Speaking of part of Derry's testimony Morrison said it showed,

... he was not at all knowledgeable with respect to neutrons and beryllium. He was also in error when he answered in the affirmative the question 'Can a scientist and can you perceive (from the testimony and the sketch) what the actual construction of the bomb was?'. (And, he was even more misleading when he answered a subsequent question 'Does the information that has been read to you, together with the sketch, concern a type of atomic bomb which was actually used by the United States of America?'. Answer: 'It does. It is the bomb we dropped at Nagasaki, similar to it.') Say rather it was a caricature of that bomb.[4]

According to Linschitz the evidence had produced a garbled, ambiguous and highly incomplete description of the plutonium bomb of 1945. He said:

... despite so many authoritative statements to the contrary by scientists over the past two decades, the layman still clings to the misconception that there is a 'secret' or key 'formula' for the construction of an atomic bomb. ...
The statement made by Judge Kaufman, when passing sentence on the Rosenbergs, regarding the technical importance of the information conveyed by Greenglass has no foundation in fact. Rather it expresses a misunderstanding of the nature of modern technology, a misunderstanding which, in this case, has had tragic consequences.[5]

13 Vladimir Kachenko (in the open-neck shirt) and his wife being escorted by British police from a Russian airliner at London airport on 16th September 1967. A policeman is fending off a Russian who is clambering onto the gangway to try to stop the couple leaving

14 Klaus Fuchs leaving for East Berlin after his release from prison

15 Bruno Pontecorvo in Gorky Street, Moscow, in 1955

The pyrotechnical finale to the 1939–45 war and the spectacular spy trials which followed, within five years of the atomic bomb, put on full view to the many who had not yet recognised the obvious signs, that scientists were men of consequence in the community. It was the influence of their work in time of war, rather than any of the many contributions they had made to the peacetime economy, which had dotted the *i*'s and crossed the *t*'s. They had also during that war, and subsequently, put on display with mixed results, their varied attempts to influence the way in which society applied this work. Not all these efforts had met with the approval of society. As a result their status, their function and their ability were still ill-defined.

Because some among them had defined their own terms of behaviour within society and had delineated their own boundaries of loyalty to the community, because they had not consulted those outside their group who had only small comprehension of their esoteric discoveries, scientists were now being looked on with suspicion by the layman. Because some had behaved as a breed unto themselves they were being treated as such. Because some had adopted a unique code of behaviour, they were being judged according to a unique code.

It was not only the men themselves who were being misunderstood, there was a dangerous lack of comprehension of their work: what Linschitz called 'a misunderstanding of the nature of modern technology'. The layman had been led mistakenly to believe that there was such a thing as *a secret* of the atomic bomb, or the hydrogen bomb, or the cobalt bomb; he had been led mistakenly to believe that scientists could choose to handle this secret as they might handle a gun: something they could use as a defensive instrument, use to threaten the rest of mankind, use for a barter, or sell to a foreign country whose creed they preferred. As science was advancing, so the scientist was finding the language of science could be translated only with difficulty for the layman. And the problem was promising to get worse rather than better.

The result of these conflicting factors on scientists themselves was to create uncertainty and self-doubt. A period of self-analysis and of recrimination was on the point of setting in.

Jacques Monod's suggestion, that those American scientists who had failed in their attempts to achieve clemency for the

S

Rosenbergs had succeeded in demonstrating the weakness of American scientists as a unified force in their own country, exemplified both the self-analysis and the recrimination. Paradoxically America was the one country of all countries in which the technocracy might have been expected to be both self-confident and influential. As a group, scientists had singularly failed to come to terms with their society. On the other hand, the rest of society had not yet learned to come to terms with science; if the layman's inability to understand the nature of modern technology had had tragic consequences in the case of the Rosenbergs, then the tragedy was not to be the last of its line. A difficult period still lay ahead. Several more individuals were yet to suffer in different ways, because of attitudes which had set in in the years leading up to the executions.

13: We Built One Frankenstein

Confused attitudes to the scientist and his role had been in existence for several years. Shortly after the Nagasaki bomb brought the Pacific war to a full-stop, Foyle's threw in London a large-scale literary luncheon party to which were invited a number of noted British academics. Now knighted for his contributions to British physics, Sir George Thomson rose there to talk about the consequences of his fellow scientists' work. He said:

> If, which God prevent, there is another war, an atomic bomb can be used. I don't believe that the number of people killed will be necessarily greater than in this war: it may well be less. . . . Now at least we know what to expect, and I believe that the knowledge will be the greatest force on the side of peace and sanity.[1]

But his remarks, designed to create optimism for the future in the mind of the layman now well aware of the tangible products of nuclear physics, met with only polite applause. At the same lunch Professor C. E. M. Joad adopted more pessimistic attitudes:

> There has, I think, run through the country an almost universal reaction of fear and horror. There was a woman I heard with a lift attendant at Hampstead saying two mornings after it was announced: 'Makes one wonder what's going to happen to one's kiddies in 20 years' time.' It does indeed!

And of the men who made the weapon, Joad had this to say: 'Will nobody ever stop the scientists. Won't somebody put them in a bag and tie them up – or into a lethal chamber before they have completed our destruction?'[2]

His comments, unlike those of Thomson, met with warm applause.

Scientists, who had worked under such great pressure and with such singlemindedness in the successful attempt to build the weapon to end the war, were being taken by surprise by the widely spread attitude of revulsion towards the result of their labours. With the possible exception of poison gas in the First World War, no other laboratory product had given rise to such a feeling of world-wide dismay.

In the months which followed the end of the war the scientists who had been cloistered for so long in the bomb laboratories were able to take stock and plan their futures. Some of these men suffered if not a revulsion, then a reaction which made them wish to dissociate themselves from the manufacture of the tools of war. Others merely wished to return to the peacetime careers from which war had diverted them. In particular, defections from the bomb laboratories by physicists of some standing in the academic world were particularly evident. As early as October 1945 Robert Oppenheimer had resigned as director of Los Alamos in order to get back to his preferred career: that of a teacher at the California Institute of Technology. He was only one of several who turned their backs on the old school house in the New Mexican mountains with a sense of both achievement and release to head for chairs of physics, chemistry and metallurgy in Princeton, Berkeley, Oxford and Cambridge.

But although there was a significant movement among scientists dissociating themselves from atomic weapon manufacture, there was no let-up in their attempts to bring home to non-scientists in general and to politicians in particular, the necessity of understanding the wider consequences of their work. In the past scientists could fairly have been accused of being politically naïve, and this they themselves were willing to acknowledge. The groups which, in the middle of 1945, had been formed at the Metallurgical Laboratory in Chicago and elsewhere to provide a platform for the ideas of social and political implications of atomic power, were now to form useful examples for more considered organisations of scientists to present concepts of how their work should benefit mankind. Moreover, it was found that there existed a few people in certain influential places who were willing not only to listen to, but even to aggrandise the scientist. Spectacular, if fearful, success in what to some laymen was still only a fractionally comprehensible field, led them to attribute omniscient qualities to the back-room boy. The physicist Samuel Allison noted the signs only too well when he, with his colleagues, was brought out front:

> Suddenly physicists were exhibited as lions at Washington tea-parties, were invited to conventions of social scientists, where their opinions on society were respectfully listened to by life-long experts in the field, attended conventions of religious orders and discoursed on theology, were asked to endorse plans for world government, and to give simplified lectures on the nucleus to Congressional committees.[3]

If scientists felt that they had failed tragically in their pre-bomb attempts to organise themselves and make their views felt, in the closing months of 1945 they rallied round to ensure that they would be ready to be heard when the future demanded it. During that period groups such as the Atomic Scientists of Chicago, the Federation of Atomic Scientists and the Federation of American Scientists blossomed like leaves from a common bud. Their concern was the international control of atomic energy, the removal of the barriers of secrecy which were so hampering both international cooperation and scientific progress, some form of domestic control for the atom, and dissemination of information in general. To this latter end a group of Chicago scientists founded the still flourishing *Bulletin of the Atomic Scientists* whose purpose, according to its editors, was an emergency undertaking designed to 'make fellow scientists aware of the new relationships between their own world of science and the world of national and international politics', and to 'help the public understand what nuclear energy and its application to war meant for mankind'.

The men who attached themselves to these self-appointed opinion and pressure groups did so because they felt profound responsibility and concern for the effect of the new technology on their country and the world. There was little of personal gain to be acquired from these and from more formal associations. Several of the more eminent scientists, and Robert Oppenheimer was one of these, found that the time they had now hoped to devote to teaching and to research was being eagerly grasped by a government badly in need of advice from those familiar with both the technology and its likely social consequences.

When in the United States in the autumn of 1945 the May–Johnson bill proposed a transition from the wartime administration to what was essentially a continuation of military supervision, Oppenheimer testified in the bill's favour. Other leading physicists such as Enrico Fermi and Ernest Lawrence also gave it their support. But the rank and file of scientists were in bitter and loud opposition at the possibility of nuclear power not being put in the hands of civilians where they believed it belonged, and at the prospect of continued secrecy in atomic matters which the May–Johnson bill threatened. Eventually, in July 1946, with a majority of scientists, including Oppenheimer, backing it, the McMahon Act was passed to give civilian control. Surprisingly, this Act survived what was for scientists a worrying black period when the revelations of the Gouzenko affair and the Canadian spy-ring raised enough hysteria to threaten more rather than less military domination over atomic scientists and their matters.

In the months following his resignation from Los Alamos Oppenheimer was only infrequently able to spend time with what he regarded as the love of his life: the teaching of physics. Increasingly his extraordinary qualities were being tapped to aid in governmental and political problems well removed from the usual domain of the theoretical scientist.

Oppenheimer had long advocated the wisdom of sharing atomic secrets – if secrets they were – with the rest of the world. When Dean Acheson, the Acting Secretary of State, appointed a special Board to outline conditions under which international cooperation on these matters could be achieved, Oppenheimer was invited to serve. Heading the committee was David Lilienthal, and the report, which it succeeded in having accepted only with difficulty, was to be known as the Acheson–Lilienthal report. There can be little doubt, however, that the person on the Board who best understood the deep issues involved, and who most strongly influenced its conclusions, was Robert Oppenheimer.

It was this report which was to serve as the basis of proposals for international control at the United Nations when its new Atomic Energy Commission convened. They were to be presented, not by a scientist, but by the symbol of Wall Street success, speculator turned statesman, 75-year-old Bernard Baruch.

Scientists disapproved. Vannevar Bush, enough of a friend to be able to speak his mind, told Baruch he was the most unqualified man in the country for the task. Baruch left a record of the conversation which the remark prompted:

'Doc,' he said, 'you couldn't be more right. Put on your hat and let's go and tell the President, and that will let me out.'
'If you get out,' said Bush, 'the damn thing will blow up.'
'If I stay in, I'm unqualified, and if I get out, it will blow up,' was the reply. 'What do you want me to do?'
'Oh, hell,' said Bush, 'stay in.'[4]

Baruch stayed in. His address to the assembled delegates of the Commission at its first session in June 1946 began with slow and deliberate rhetoric: the voice of aged and experienced authority:

> We are here to make a choice between the quick and the dead. That is our business.
> Behind the black portent of the new atomic age lies a hope which seized upon with faith, can work our salvation. If we fail, then we have damned every man to be the slave of Fear. Let us not deceive ourselves: We must elect World Peace or World Destruction.
> Science has torn from nature a secret so vast in its potentialities that our minds cower from the terror it creates. Yet terror is not enough to inhibit the use of the atomic bomb.

The choice he and his colleagues favoured was international agreement to cease the manufacture of atomic bombs; disposal of existing bombs; an inspection plan; and the formation of an international atomic development agency. But whatever hopes were entertained for Russia settling immediately for the quick rather than for the dead rapidly evaporated; this was only the beginning of the haggling over atomic control which was to go on and on for years.

Less than three weeks after Baruch had attempted with his oratory to neuter the owners and would-be owners of atomic weapons, the United States began a series of atomic tests at Bikini Atoll. Grave fears had been expressed about the possible consequences of these underwater nuclear explosions. Some believed that the result might be the annihilation of under-water life in a vast area of the Pacific. Others prophesied the initiation of a vast chain reaction in the waters of the ocean which might spread in a catastrophic fashion over the surface of the earth. In the end, however, the worst effect was the predictable reaction from Russia. America, it seemed to the Soviets, was behaving like an abolitionist hangman holding a dangling noose: she was proposing a halt to atomic weapon development at the same time as she was refining her technique.

Still the strong reaction of some scientists against continued weapon development and testing had continued, not only in the new organised quasi-political groups, but also from articulate and highly vocal individuals. There had even been letters from scientists to the *New York Times* suggesting a United States declaration of policy that bomb production should stop and stockpiles be destroyed.

But by no means all scientists who knew the issues involved were convinced that this rush to disarm and to dissociate from weapon development was a commendable attitude. To many there were other urgent considerations: and some of these were tempered with fear. There were questions to be asked which only men familiar with the likely rate of international scientific advance could begin to answer. The first obvious question was how far had other nations' scientists progressed in atomic weapon development? But there were several others; for example, were the uranium and the plutonium bombs really the ultimate in destructive devices which the devastation of Hiroshima and Nagasaki suggested? Was there a quantum jump in scientific weaponry yet to come? And who would own it if it did come?

When war ended, there was no slackening of interest in the abilities of the other competitors in the atomic bomb race; guessing the date when the runner-up would breast the tape was a common enough pastime among scientists, politicians and military men. But only the time when the contestant would arrive, rather than his identity, was at serious issue. Germany, once the favoured though ill-judged candidate, had long been crippled. Britain was delayed by handicap; its close association with the United States in the Manhattan Project gave it great initial advantages but the McMahon Act had prevented any further military cooperation: Clement Attlee was later to call it 'that stupid McMahon Act'. Russia with great resources of men and materials was bound to arrive soon. But how soon?

In 1945 estimates varied widely. Some scientists suggested what was then the almost inconceivably short time of five years; others, General Leslie Groves among them, believed that Russia might be capable of producing an A-bomb before 1965. But as the years passed, as more and more Russian scientific expertise was conceded, and as the atomic spy trials mounted in their frequency, the longer estimates fell away and 1950 began no longer to seem unrealistic as the date to expect the first Soviet atomic tremor.

However, Russian parity in atomic arms was not the limit of a few scientists' concern. Even whilst they had been working on their new weapons, Manhattan Project physicists had been well aware that theory did not preclude the construction of a device whose effects could dwarf those of the uranium and plutonium bombs. If scientists in the West were alert to the potential existence of this frightening prospect, then Russian scientists must by now be equally alert. The more surely since the fundamental principles of this, the hydrogen bomb, were known to have been under discussion in Russia in the 1930's. Indeed, the man who had been responsible for inspiring some of the initial theoretical work which led to thoughts about this weapon in the West was a refugee to the United States from Russia.

George Gamow was a freshly graduated physicist from the University of Leningrad who was lucky enough to further his trade on short visits abroad. A bubbling extrovert, he soon took a liking to the Western way of life and the atmosphere of unbounded intellectual freedom which brief personal contact with leading physicists such as Rutherford and Bohr gave him. During his travels Gamow became interested in the tentative theories which a few European scientists were proposing to explain the mechanism by which the energy of stars was produced. The suggested process was that small atomic nuclei, such as those of hydrogen, could be made to fuse together with the resulting emission of enormous amounts of energy. Because such reactions could take place only under conditions of exceedingly high temperature, they were to become known as 'thermonuclear' reactions.

On his return to Russia Gamow frequently discussed these new theories and in 1931 shortly after his election to the Soviet Academy of Sciences, at the tender age of 27, he addressed the Academy on the subject. Present at this meeting was a Soviet official named Bukharin, to whom had been delegated development of national scientific and technical resources. Impressed and inspired by what he had just heard, Bukharin offered to put the electricity output of Moscow or Leningrad completely at the physicist's disposal. For an hour or so by night, if necessary halting all domestic and factory supplies, Bukharin suggested that Gamow should pass a massive current through a thick wire to produce the high temperature. Gamow could merely politely refuse the offer and explain that this method would not be capable of getting the temperatures he had in mind: perhaps several hundred million degrees. Even today a controlled thermonuclear reaction as a large-scale utilisable energy source is still a physicist's dream.

Gamow escaped from Russia in 1933 and it was Bukharin who helped him negotiate permission for his wife to leave the country. When he arrived in the United States a year later Gamow read newspaper reports of Bukharin's execution by firing squad.

In America Gamow found many others as intrigued as he by the theoretical aspects of his problem. It was in 1938 that Hans Bethe, who a few years earlier had left Germany during the Nazi persecutions, produced a satisfactory theory to explain the fusion reactions responsible for stellar energy. Edward Teller, the Hungarian who had joined Gamow in the department of physics at George Washington University, was another whose curiosity had been stimulated. The state of the art was then what Teller called 'a game, an intellectual exercise' of thermonuclear physics.

But this toying with a heavenly problem was to remain a game for only a few years. As a prospect very much of this earth, thermonuclear reactions as a basis for a super bomb were under detailed consideration in the year in which America entered the war. Even before Los Alamos was chosen as the site where the uranium fission weapon could be assembled, those who were to make key contributions to the product, Oppenheimer, Bethe, Teller, Fermi and many others, were discussing how the successful fusion of atomic nuclei might be achieved in theory and hurriedly engineered into practice.

If the uranium fission bomb could be made to work, then here on earth would be a vehicle for the high temperature required to trigger a fusion reaction. One of the first of the prefabricated buildings to be thrown up in the rush to put Los Alamos on an experimental footing was a construction designed to handle materials for thermonuclear work. An investigation of the properties of tritium, the heavy isotope of hydrogen, for which a key role in the thermonuclear bomb was anticipated, was one of the earliest of the projects to get underway.

The responsibility for the initiation of this work lay with Robert Oppenheimer as did the responsibility for the whole of the

scientific administration of the laboratory. His brief was to produce a massive weapon. The crucial direction in which he aimed
the work of his fellow physicists would spell out success or failure
in terms of the initial brief. Inevitably, in such an unprecedented
scientific undertaking, his personal success or failure would be
linked to this direction, for the stakes were enormous.

The dream of thermonuclear power, like the dream of nuclear
power, was conceived by physicists in time of peace. In both cases
the concerted efforts of the most able practitioners of scientific
theory and practice were to realise the dream. In both cases war,
or the threat of war, was to be the stimulus, without which progress towards the desired end would have taken years and years.
And in both cases the desired end was a bomb. But in 1943, when
there was no means whatever of guaranteeing success, it was clear
that commitment could be to one large-scale project only.

Oppenheimer's resources were vast but finite. He committed
them to the atomic bomb and to success. War ended with the
thermonuclear weapon as nothing more than a recognisable and
mercifully unneeded prospect.

It was this prospect on which many of the leading scientists of
the bomb laboratories turned their backs in 1945 to face a peacetime life. The cost of the war in terms of human life and misery
had been totted up; the cost of Hiroshima and Nagasaki in terms
of human life and misery had also been totted up and proved to be
small by comparison with the rest of the world-wide holocaust.
It was in spite of this arithmetic that, when war ended, and the
stimulus to dedicate science to war vanished, many scientists now
looked at the results of their work and experienced from it a
profound moral repugnance. The future ought not to have need
of a thermonuclear bomb.

Less than a year after Oppenheimer moved into Los Alamos,
Captain Peer de Silva, one of the officers charged with maintain-
ing the military security of the precious establishment, wrote a
memorandum to a senior officer in which he said:

> It is the opinion of this officer that Oppenheimer is deeply con-
> cerned with gaining a worldwide reputation as a scientist, and a
> place in history, as a result of the D.S.M. [Manhattan] project. It is
> also believed that the Army is in the position of being able to allow
> him to do so or to destroy his name, reputation, and career, if it

should choose to do so. Such a possibility, if strongly presented to him, would possibly give him a different view of his position with respect to the Army, which has been, heretofore, one in which he has been dominant because of his supposed essentiality. If his attitude should be changed by such an action, a more wholesome and loyal attitude might, in turn, be injected into the lower echelon of employees.[5]

Clearly, the young military man was disturbed that the scientist's god was not the same god to which he himself owed homage. And clearly his sweeping opinions were coloured by disenchanting clashes between his own personality and those of scientists in general, and Oppenheimer in particular.

It was inevitable in the heterogeneous community of Los Alamos, consisting as it did of scientists and engineers on the one hand, and the military on the other, that apparent motives and allegiances should be in opposition. It was equally inevitable, though this must have been less easy for the military to understand, that there should be conflicts in aims and priorities among the leading scientists themselves.

Those periods in time which led to the birth of the thermonuclear weapon were seldom free from such conflict. At Los Alamos Oppenheimer initially encouraged thermonuclear research and the work of the man who was to emerge as its devoted champion, Edward Teller. But Teller has never chosen to hide the fact that he was bitterly disappointed that Oppenheimer should eventually have allowed thermonuclear work at Los Alamos to slip into a backwater.

However, there was more than a conflict merely of aims between the two men. Teller, the ebullient and over-forthright Hungarian, a man of great fixity of purpose, inspired devoted followers in the scientific field, but antagonised others. Some Los Alamos workers observed that whilst Oppenheimer outwardly admired Teller's obvious talents as a physicist, he rejected any close bond of personal friendship.

In the weeks following the successful explosion at Alamogordo and preceding the end of the war, interest in the fusion weapon revived, but as quickly died again. It was in the few days between the dropping of the Hiroshima and Nagasaki bombs that Oppenheimer told Teller he would have nothing more to do with thermonuclear work.

268

There were few among Teller's colleagues at that watershed of warfare who could conceive that a case might be argued to devote physics to yet another weapon. Science had been drawn into the last conflagration of nations without reluctance; but it was not to be drawn easily again now that the conflagration had ceased.

For four years, whilst the United States stockpiled a nuclear armoury, there was a paucity of interest in thermonuclear weapon research. Then the incident occurred which yet again was to abut the forefront of scientific knowledge with war. In August 1949 a United States B-29 bomber returned from a scientific observation mission in the Far East carrying a series of photographic plates. On development these plates showed unmistakable evidence of a recent atomic bomb explosion. U.S.A.F. and R.A.F. detector aircraft quickly confirmed that Russia had exploded a nuclear weapon. President Truman made the announcement of the intimidating news on 23rd September.

The factors which had dissuaded Western scientists from developing new weapons of great magnitude had had no influence with their Russian counterparts. Had any doubts remained of the rapid post-war progress of Soviet nuclear technology, then this more than adequate demonstration removed them. And no doubts need be tolerated that whatever progress had been made towards a thermonuclear weapon in the United States would ultimately be matched in the Soviet Union. But was this enough reason for scientists to compete?

It was still less than 10 years since nuclear physicists had been swept into wartime careers as bomb-makers. Still the same generation of men recruited in Hitler's war remained; still their singular skills would be required to meet Russian competition. For the second time in a lifetime they must face commitment of their science to the war machine. But many were still unsure of the attitudes of the rest of mankind to their last creation. What problems had their intervention in the last war solved? An initial approbation had given way to revulsion. For all its potential as a force for peace, failure to achieve international control of the mighty atom had let it remain as a threat of war. And what now? If thermonuclear weapons could be quickly built and so raise the level at which war could destroy by yet another order of magnitude, what questions would this settle?

There were divided attitudes among scientists. Hans Bethe was only one of many who, during the next few years, gave much of his time and thought to the problems raised by the thermonuclear weapon. He described his contributions as those of a small cog in a big wheel; a quiet-spoken, perhaps over-modest man, he was nonetheless a much respected and influential physicist, and the troubles in his mind were the troubles of many of his colleagues during that period.

During the past few months Bethe had been giving some advice on minor applications of thermonuclear principles to the Los Alamos laboratory, and the news of the Russian bomb now deeply worried him. Shortly after the news broke, Teller had appeared on the campus of Cornell University where Bethe had taken a professorship, and the former Los Alamos colleagues had discussed the new situation.

Bethe was impressed by Teller. Teller had by now developed both new technical ideas which might contribute significantly to a thermonuclear programme, and a deep conviction that the new emergency demanded unrestrained action by the best available scientific minds. He wanted Bethe to return full-time to Los Alamos to give his considerable talents as a theoretical physicist to a crash super-weapon programme to meet the Russian threat.

Bethe straddled a fence of uncertainty. The technical prospect of the new work was tempting, the political argument in favour of it was strong, but the moral justification required to support it was troublesome. He turned with Teller to Oppenheimer, now director of the Institute for Advanced Study at Princeton. Oppenheimer, he found, when they met, was no less troubled by the situation, but strangely obscure. This attitude worried Bethe. Between him and Oppenheimer there existed a strong bond of mutual respect. Bethe had learnt to expect considered guidance from his wartime leader; on this occasion he was not sure what he was getting. During the discussion Oppenheimer quoted James Conant as having said that he could never be expected to approve the development of a weapon a thousand times more powerful than the atomic bomb, but Bethe was unable to make out whether Oppenheimer agreed with Conant or not.

Teller made clear to Oppenheimer that he believed that now was the time to go all out for the new weapon. Together, and to Bethe's surprise, they began to discuss what advantages might come from scientific cooperation *with* Russia on thermonuclear research. To Bethe the idea made no sense at all, yet he found that the other two agreed that this was a proposition worth considering.

Bethe, when he left Oppenheimer, had still not eased what he called his 'inner troubles' and was no nearer to deciding whether or not he should join Teller on the super-bomb. He turned now to other physicist colleagues. At the end of one long discussion Bethe and Victor Weisskopf had to agree that there was already enough evidence to show that a war fought with hydrogen bombs would result in a world not worth preserving; even if they were on the winning side they would share in the responsibility for destroying the values they were seeking to preserve. Weisskopf's committed attitude opposing thermonuclear weapon research caused Bethe to commit his own; he telephoned Teller to tell him that he could not work on the new project.

A term of trial for the scientific conscience was again in full swing. The spectrum of commitment by Teller, indecision by Oppenheimer, rejection by Bethe typified the conflict among U.S. nuclear physicists as a whole. But Oppenheimer had taken on greater responsibilities than most. He had accepted the chairmanship of the General Advisory Committee (G.A.C.) of scientists to advise the Atomic Energy Commission led by David Lilienthal.

During the period when scientists were so earnestly discussing and trying to align their attitudes, Oppenheimer was asked to convene his committee to advise the Commission on the thermonuclear weapon. Before this meeting and shortly before Bethe had visited him he had written to James Conant, also a member of the G.A.C., with some background information which he felt he would like 'Uncle Jim' to consider before the committee gathered. In his letter Oppenheimer had said:

> . . . a very great change has taken place in the climate of opinion. On the one hand, two experienced promoters have been at work, i.e. Ernest Lawrence and Edward Teller. The project has long been dear to Teller's heart; and Ernest has convinced himself that we must learn from Operation Joe [the Russian atomic bomb] that the Russians will soon do the super, and that we had better beat them to it. . . .
> The climate of opinion among the competent physicists also shows signs of shifting. Bethe, for instance, is seriously considering return on a full time basis; and so surely are some others. I have had long talks with Bradbury [Oppenheimer's successor at Los

Alamos] and Manley, and with von Neumann. Bethe, Teller, McCormack, and Le Baron are all scheduled to turn up within the next 36 hours. I have agreed that if there is a conference on the super program at Los Alamos, I will make it my business to attend.

What concerns me is really not the technical problem. I am not sure the miserable thing will work, nor that it can be gotten to a target except by ox cart. It seems likely to me even further to worsen the unbalance of our present war plans. What does worry me is that this thing appears to have caught the imagination, both of the congressional and of military people, as the answer to the problem posed by the Russian advance. It would be folly to oppose the exploration of this weapon. We have always known it had to be done; and it does have to be done, though it appears to be singularly proof against any form of experimental approach. But that we become committed to it as the way to save the country and the peace appears to me full of dangers.[6]

On the day following his Atomic Energy Commission's meeting with Oppenheimer's advisory committee, on 29th October 1949, David Lilienthal made some notes for his journal which show the unease with which the scientists gathered about the discussion table to face the hydrogen bomb:

Conant flatly against it. Hartley Rowe, with him: 'We built one Frankenstein.' Obviously Oppenheimer inclined that way. Buckley sees no diff. in moral question x and y times x, but Conant disagreed – there are grades of morality. Rabi completely on other side. Fermi, his careful enunciation, dark eyes, thinks one must explore it and do it and that doesn't foreclose the question: should it be made use of? Rabi says decision to go ahead will be made; only question is who will be willing to join in it.'[7]

These were men who were carrying a fearsome responsibility on behalf of their country. The moral problems were enormous and the consequences for civilisation as a whole potentially crucial. The advice of those present was eventually given as a unanimous statement rejecting the thermonuclear programme. The G.A.C.'s recommendations included the words:

We all hope that by one means or another, the development of these weapons can be avoided. We are all reluctant to see the United States take the initiative in precipitating this development. We are all agreed that it would be wrong at the present moment to commit ourselves to an all-out effort towards its development.[8]

Chairman: Gordon Gray   Ward V. Evans   Thomas A. Morgan

16 Members of the Personnel Security Board of the United States
Atomic Energy Commission, in the matter of J. Robert Oppenheimer

17 President Kennedy presents the Enrico Fermi Award to Edward Teller at the White House, 3rd December 1962

18 A vaccine production unit working at the Microbiological Research Establishment at Porton Down, Wiltshire

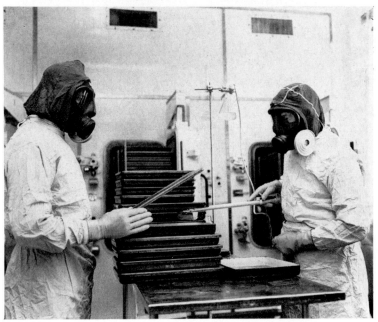

To the main recommendations Enrico Fermi and Isidor Rabi added a minority report which said:

> The fact that no limits exist to the destructiveness of this weapon makes its very existence and the knowledge of its construction a danger to humanity as a whole. It is necessarily an evil thing considered in any light. For these reasons, we believe it important for the President of the United States to tell the American public and the world that we think [it] is wrong on fundamental ethical principles to initiate the development of such a weapon.[9]

The scientists' advice to the President was clear and unified. The Atomic Energy Commission approved its advisory committee's recommendations and suggested that President Truman should do the same. But the Commission was not unanimous. One of its members, Lewis Strauss, a man who once described himself as 'a Black Hoover Republican', had long believed that a new order of nuclear weapons was the only hope of guaranteeing America's survival and he wrote to Truman recommending that the Atomic Energy Commission should 'proceed with all possible expedition to develop the thermonuclear weapon'. His was by no means the only lobby in opposition. Some physicists with very considerable reputations had been making noises in direct contradiction to the unanimous views of the highly selective advisory group of scientists. Lilienthal later described how Ernest Lawrence and Luis Alvarez had spent some time with him 'drooling' over a super weapon capable of devastating vast areas. Lawrence, Alvarez and Teller had put impassioned pleas for the weapon's right to existence to the Joint Committee on Atomic Energy, and influential members of government had been won over.

Truman was clearly unimpressed by the conclusions of the committee of scientific minds convened especially for the purpose of giving him advice. On 31st January 1950 he broke the news to the nation that work would 'continue' on all forms of atomic weapons including the 'so-called hydrogen or super bomb'. Political and military dictates had outweighed scientific caution.

On 3rd February another announcement was made which was of more than passing interest for Truman: Klaus Fuchs had been arrested in London. Fuchs was known to have attended some of the early thermonuclear conferences at Los Alamos.

T

For those who were convinced that work on the super weapon was a justifiable métier, the path to Los Alamos was open. The laboratory had a new purpose. In the months that followed some of the most ingenious of American nuclear physicists – the leader, Norris Bradbury, Edward Teller, Stanislaw Ulam, John Wheeler, John von Neumann, and many others – made their contributions.

But those with moral qualms had as yet had no reason to reverse them, just as those who opposed the thermonuclear weapon on its military and political undesirability had as yet had no reason to suspend their opinions. In an article published in April 1950, Hans Bethe made public the misgivings of his conscience.

> If we fight a war and win it with H-bombs, what history will remember is not the ideals we were fighting for but the methods we used to accomplish them. These methods will be compared to the warfare of Genghis Khan, who ruthlessly killed every last inhabitant of Persia.[10]

He ended his article with a plea for international control of atomic weapons and the outlawing of 'the greatest menace to civilisation, the hydrogen bomb'.

But his vacillations were by no means at an end. In less than three more months he was to offer his services to the Los Alamos thermonuclear programme. The outbreak of the Korean war up-ended his resolve.

To him as to many others who felt bound to involve themselves in weapon research, the reality of war lowered the price of commitment. This, nevertheless, did not make the work's purpose any less distasteful. Bethe even harboured the wish that his theoretical contributions would prove that a fission bomb was a scientific impossibility and make mockery of the efforts of both American and Soviet physicists, so ending the problems of politicians, tacticians and scientific moralists.

For many months of 1950 and 1951 the challenge provided by the new weapon looked formidable. The initial optimism with which Los Alamos had swung itself into its new purpose began to wane in the face of unforeseen problems of science and engineering.

Then in May of 1951 a successful test explosion at the Eniwetok Atoll in the Pacific Ocean raised hopes. But even so many physicists were still convinced that the only significant question of that time was not whether the hydrogen bomb should be built, but whether it could be built. The A.E.C. therefore decided that the time had come to gather together all the men who might make contributions to the fusion weapon, hand them chalk and a blackboard, and have them draw up the best line of attack.

Oppenheimer as chairman of the Weapons Committee of the G.A.C. was asked to preside at a meeting at his own Institute on 16th June. Gordon Dean, the new chairman of the A.E.C., later described how on that occasion there were present at Princeton,

> . . . every person, I think, that could conceivably have made a contribution. People like Norris Bradbury, head of the Los Alamos laboratory, and 1 or 2 of his assistants, Dr Nordheim, I believe was there from Los Alamos very active in the H program. Johnny von Neumann from Princeton, one of the best weapons men in the world, Dr Teller, Dr Bethe, Dr Fermi, Johnny Wheeler, all the top men from every laboratory, sat around this table and we went at it for 2 days.[11]

Recollections as to what took place around that table differ. According to Edward Teller, a vital hydrogen bomb report he had submitted three months before was not considered and many scientists and officials spoke without so much as a mention of it until, in anger, he insisted on his ideas being heard; only when he carried out calculations on the blackboard did he begin to get any response for his new way of approaching the thermonuclear weapon. According to Hans Bethe and others, Teller's proposals were well understood before the meeting and were discussed almost from the start. That Teller did at some point get a warm and enthusiastic response is, however, not in dispute, and it is more than likely that Teller's dynamic oratory and powerful abilities to persuade – which once experienced are not easily forgotten – washed a spirit of great optimism and purpose over the meeting.

From when he first knew of it, Bethe had rated Teller's contribution very highly:

> It was one of the discoveries for which you cannot plan, one of the discoveries like the discovery of the relativity theory, although I don't want to compare the two in importance. But something which is a stroke of genius, which does not occur in the normal development of ideas. But somebody has to suddenly have an inspiration. It was such an inspiration which Dr Teller had . . . which put the program on a sound basis.[12]

There and then this select group of American physicists threw themselves into the calculations which at last were to lead to a feasible method of producing the H-bomb. They included those who not many months before had represented every portion of the spectrum of moral and political opinion on the thermonuclear weapon: Teller, Oppenheimer, Bethe, Fermi. The meeting closed with every scientist enthusiastic and convinced that it had a future. The chairman, Oppenheimer, according to Bethe, 'entirely and wholeheartedly supported the programme'.

The last leg had been reached. But though the technical hurdles had been lowered still others remained in the months ahead. Teller let it be known that he was dissatisfied with the administration of the Los Alamos thermonuclear programme and even as late as the autumn of 1951 unsuccessfully tried to persuade Oppenheimer actively to join the project. Eventually, in November 1951, Teller himself parted company with Los Alamos. But the way was now clear for what he – and many other physicists with him – believed to be the unavoidable and dutiful creation of science.

On 1st November 1952, in a shed on the small island of Elugelab in the South Pacific, the device was prepared: the result of months of bitter political wrangling and the culmination of years of work of dozens of physicists. The thermonuclear weapon flowered into the greatest explosion on the face of the earth.

Elugelab disappeared, leaving behind a crater into which, some cynical mathematician calculated, would fit fourteen Pentagon buildings.

Ten months later Premier Malenkov announced to the Supreme Soviet that the U.S.S.R. could again match the United States, weapon for weapon. Within weeks of testing his country's first atomic bomb in 1949, Igor Kurchatov had set about the thermonuclear problem, and 'before sunrise on 12th August 1953' exploded a hydrogen bomb 'exactly as planned'.

14: The Extraordinary Personage

On 24th July 1946, David Lilienthal, who in a few weeks was to
accept the chairmanship of the United States Atomic Energy
Commission, sat waiting at Washington airport and added a few
random thoughts to the personal journal he had been keeping for
many years. He began the account of the past 24 hours with some
notes about J. Robert Oppenheimer:

> Met J.R.O. last night, just in from New York; talked until 1.30
> this morning. He is really a tragic figure; with all his great
> attractiveness, brilliance of mind. As I left him he looked sad:
> 'I am ready to go anywhere and do anything, but I am bankrupt
> of further ideas. And I find that physics and the teaching of
> physics, which is my life, now seems irrelevant.' It was this last
> that really wrung my heart. Here is the making of great drama;
> indeed, this *is* great drama.[1]

With prescience which in later years must have surprised him,
Lilienthal had foreseen some catastrophe ready to enter the life
of the physicist. Others had seen only the outward unusual
characteristics of the man: and these were impressive enough. He
was, as Lilienthal told his journal after their first meeting, 'an
extraordinary personage': a scientist of apparently measurably
high success of whom Harry Truman wrote, 'More than any other
man, Oppenheimer is to be credited with the achievement of the
completed Bomb'. When the time came for his wartime colleagues
to give their testimonials they were to use mainly superlatives in
their descriptions of Oppenheimer and of his abilities. Vannevar
Bush described him as 'one of the great physicists of this country
or of the world for that matter'. Of his powers as an orator one

277

professor of chemistry said, 'I have seen him sway audiences. It was just marvellous, the phraseology and the influence is just tremendous. I can't analyse it for you. . . .' The same man stated his belief that many young scientists leaving Los Alamos after the war left as pacifists, such was the power of Oppenheimer's influence over them.

Middle-aged men, who were not much more than boys when they first came under Oppenheimer's influence as wartime leader of the Manhattan project, today still reverently guard their memories of him as teacher, mentor and inspirational guide, whilst older men, among them some of the world's most senior and distinguished physicists, still readily acknowledge the value of the experience of working alongside him, and of the time spent in his company.

Even the man's physical characteristics were unusual enough to cause some of those who met him to feel privileged; one friend of his early manhood described his appearance as that of 'a young Einstein, and at the same time like an overgrown choir-boy. There was something both subtly wise and terribly innocent about his face.' 'What about his character?' one senior physicist was once asked under oath. 'His character?' he replied. 'Ethical, moral is first rate.'

About the man who at the age of 39 had taken charge of the atomic bomb project and seen it through to its required successful conclusion in 1945, a myth had to grow. His great ability was to be seen as infallibility and his high humanity as deity: 'Among scientists Oppie is God . . .', read the headline of one popular science article.

The time came, inevitably enough, when Oppenheimer had to be exposed as a man, and, as such, to have failings. That this exposure should be carried out in the fashion it was, so as to confirm Lilienthal's prediction of great drama, and with such thoroughness as to carry Oppenheimer's life to the brink of tragedy, could never have been dreamed of in 1945, even by those least impressed by his spotless reputation. Nor could it have been guessed, particularly by scientists themselves, that in this painful process, great scars would appear on the face of unity and brotherhood which they had once presented to the world as the untroubled image of science itself.

For four years until June 1953, William Borden had been executive director of the Joint Committee on Atomic Energy. But after resigning his post Borden considered he had not yet fully discharged all his duties. On 7th November he addressed a letter to J. Edgar Hoover, the Director of the F.B.I. To describe its contents as sensational would be understatement. It began, 'This letter concerns J. Robert Oppenheimer'. It went on to make many accusations against the physicist including, that more probably than not he was an agent of the Soviet Union, that he had binding ties with communism, and that he was influential in retarding and suspending the United States H-bomb programme.

On 3rd December President Eisenhower had a message telephoned from the White House to summon Lewis Strauss, now Chairman of the Atomic Energy Commission, to an urgent meeting to discuss the matter of J. Robert Oppenheimer. The meeting ended by the President directing that Oppenheimer's security clearance to secret information should be withdrawn pending a hearing. When Oppenheimer received the news he was in England delivering the Reith lectures on 'Science and the Common Understanding'.

By 21st December he had returned to his own country and was in Washington in Strauss's office to meet the A.E.C.'s chairman and its General Manager, General K. D. Nichols. There he was given to read a draft letter by Nichols in which a number of charges were drawn up, and in which he was offered a hearing by a personnel security board of the A.E.C. There was an alternative to accepting the hearing, and this Oppenheimer was aware of: he could resign as consultant to the A.E.C. and leave in abeyance a string of accusations which were as formidable in their outlines as those which had faced any of the atomic spies convicted during the last decade.

Oppenheimer was not the first scientist to feel the rough edge of the anti-communist fervour which had allowed tolerance of the methods of McCarthyism to thrive in American society. But the accusations pierced more deeply than those which others had found tearing at the surface of their lives and their livelihoods. The very foundations of his magnificent edifice of a career based on his loyalty to his country were being attacked; his life of dedication of his science to his country's war, which had turned him into a national figure, even a national hero, was being

279

challenged as a great lie. If he refused a hearing, the accusations against him would remain on the record.

It was half a century since Oscar Wilde had been faced with a slip of paper with its misspelt message from the Marquis of Queensberry. Wilde had known that he could either leave the accusation recorded as it was, or risk the chance of public degradation; he resolved his dilemma by electing to expose himself to a court of law. Oppenheimer, though believing himself to be innocent, now found himself holding a letter which placed him in a not dissimilar dilemma. He subsequently accepted the hearing.

The Personnel Security Board began its sittings on 12th April 1954 in some bleak temporary wartime offices in Washington. Its purpose was to decide whether or not to continue the suspension of Robert Oppenheimer's security clearance.

The Board had three members: Gordon Gray, once a lawyer, newspaper publisher, Secretary of the Army, and now President of the University of North Carolina, as chairman; Thomas Morgan, former President of the Sperry Corporation; and Dr Ward Evans, Professor of Chemistry at Loyola University, Chicago, the only scientist member.

In the next weeks these three senior citizens were to expose themselves to the most initmate details in the life-story of Robert Oppenheimer. The man of their concern who faced them on that first morning was only a few days removed from his fiftieth birthday. The thick, uncontrolled shock of black hair which had once invited the comparison with Einstein had now gone, and in its place was a short, greying crop. He was a thin unaggressive figure who stooped a little, more often than not carrying a lighted pipe or cigarette. His words were slow, deliberate and long-considered before being uttered in a too quiet, sometimes indistinct voice. The trademark of each sentence was an 'er' or an 'um' which helped him the better to construct his thoughts. Frequently during the hearings he had to be asked by counsel to repeat partly heard words and phrases.

The affair was not intended as a trial; it was an inquiry, and on more than one occasion Gordon Gray held up the proceedings to remind those who were present of the fact. But form and substance nevertheless were at great variance. Both Oppenheimer and the Board had engaged top-flight attorneys as counsels and the attitudes adopted by these men from the first days gave the room the atmosphere and significance of a court of law. The Board had only one relatively insignificant punishment it could award: the refusal of access to secret information. The fact that, in the few years past, Oppenheimer had in any case been disentangling himself from government work, merely emphasised the inadequacy of the Board's power. This punishment, if awarded, could be nothing more than a symbol; the reality would be the desolation of Oppenheimer's lifetime career. The attack and defence of this career during the month that followed were to be as bitter in their methods and as cruel in their exposures as in any trial.

The proceedings began by the reading of the letter which Kenneth Nichols, the General Manager of the A.E.C., had shown to Oppenheimer in Lewis Strauss's office. It differed widely in many points from the original letter which William Borden had sent to the F.B.I., but it was damning enough for all that.

This letter, addressed to Oppenheimer, listed why his employment by the A.E.C. was no longer considered to be in the national interest. The accusations fell into two categories: first, that his connections with communism made him a security risk, and, secondly, that he had actively jeopardised the hydrogen bomb programme.

A whole string of communist front organisations was read out to which Oppenheimer had given his support in the late 1930's; he was quoted as having once said that he had probably belonged to every communist front organisation on the west coast. Short of active membership by him, the list of personal connections with the party which the letter gave could scarcely have been more formidable; they included Oppenheimer's wife, his mistress and his brother, all named as one-time members.

It was stated that his rejection of the thermonuclear weapon had been based on moral, practical and political grounds, and that even after it became a matter of national policy he continued to oppose it and dissuaded other outstanding scientists from associating themselves with its development. To Oppenheimer was allotted the title, 'The most experienced, the most powerful, and most effective member' of the opposition which organised itself against the hydrogen bomb. It was less than 10 years since these same phrases had been used to describe the same man's positive contributions to the team which built the atomic bomb.

The accusations linking Oppenheimer with communism presented in outline a simple problem for the Board. If the physicist by his associations was an unwitting security risk, as Nichols's letter in its more lenient interpretation suggested, or if he was a traitor, as Borden's original letter stated, then of course his services as adviser to a government engaged in preparations for war should be disposed of. If not, then there was, in theory at least, no case to answer. The hydrogen bomb story posed more delicate problems for the Board, and for science as a whole. It had first to be judged whether this physicist of recognised ability had presented to his government, and to those associated with him on his weapon work for that government, 'bad' advice based on moral, practical and political grounds. If this was the case, it must then be decided for what reasons this bad advice had been given. And if it had been given in good faith, did it now qualify Oppenheimer for dismissal and even disgrace?

But Oppenheimer's role as a government scientist was far from being unique. Was every responsible scientist expected always to agree with every piece of weapon policy of his government? And if not, should he then be removed from all other aspects of weapon development to which he had committed himself, merely because in one outstanding case his views had clashed with those of the majority?

These hearings had more than the career of one scientist at stake. They were to put in question the role of scientists as a whole in peace and war, examine their loyalties and, in the end, put on trial the system of security to which scientists had submitted in the past, and to which they still were submitting themselves.

Immediately Gordon Gray finished reading Nichols's letter he turned to the one Oppenheimer had sent in reply. It was an unusual document written in great detail and at great length and with the considered eloquence with which Oppenheimer had so often impressed and powerfully influenced scientists and politicians alike. It was an autobiography showing what he considered were his achievements, his loves, his hates and his motives.

In a style giving the periods it described an air almost of unreality, it told of the Jewish child born to a family which had him benefit from the priviliges of wealth. The young Oppenheimer was to pass through the best American and European universities on his way to what promised to be an untroubled life of academic physics. But the events of the late 'twenties and 'thirties, in which human lives were contorted by economics and politics, were to force him into living with reality.

The depression, the sufferings of Jews in Germany and the Spanish Civil War, all encouraged him to attune his political consciousness. He attached himself to the left-wing cause, serving on committees to provide aid to the Spanish Loyalists. As with many other of his west coast academic friends, his interest in communism ripened and he quickly aligned himself with the party's humanitarian aims. But this love-affair was brief. Two factors contributed to the beginnings of a disenchantment with communism. The first was the confirmation, by physicists he could trust who had lived through the Russian purges of 1936-9, of the rumours he had heard of the tyranny of the political trials, and the treatment meted out to the victims. The second was the Nazi-Soviet pact which apparently attempted to unite his new-found love of communism with his new-found hate of fascism. By the time he was offered Los Alamos and the role that was to change mankind's future as well as his own, he considered communist fellow-travelling to be a part of his past.

The portion of his letter which dealt with the war years was told with suitably muted pride: the pride of a man who had been stretched to his limits and who had emerged as a success, and as an international figure into the bargain.

> . . . I had become widely regarded as a principal author or inventor of the atomic bomb, more widely, I well knew, than the facts warranted. In a modest way I had become a kind of public personage. I was deluged as I have been ever since with requests to lecture, and to take part in numerous scientific activities and public affairs. Most of these I did not accept.[2]

What he did accept were numerous posts as adviser on United States nuclear policy. It was in one of these, as chairman of the General Advisory Committee, that he joined in the Committee's unanimous opposition to the development of the hydrogen bomb. Oppenheimer's claim was that when, in spite of this opposition, President Truman announced his decision to proceed with the thermonuclear programme, his own opposition ended 'once and for all'. Never, he said, had he urged anyone not to work on the hydrogen bomb project.

The letter ended,

... I have reviewed two decades of my life. I have recalled instances where I acted unwisely. What I have hoped was, not that I could wholly avoid error, but that I might learn from it. What I have learned has, I think, made me more fit to serve my country.[3]

The Board heard at length the evidence against Oppenheimer and his repudiation of the charges. To follow were many long days of searching cross-examination of Oppenheimer and his former colleagues. Frequently Oppenheimer was to find himself in difficulties as a result of the combined skill and acidity of the Board's counsel, Roger Robb. But as more witnesses passed in front of it, the Board, notably the chemist Ward Evans, showed evidence that it too was troubled. On several occasions Evans felt he should remind witnesses that the Board's role as judge was not of its own choosing. One of the witnesses, forthright Vannevar Bush, took the three men to task for having accepted Nichols's letter of charges at all; for these were capable of being interpreted as pillorying a man because of his opinions. Bush had rightly seen that the hearings were the 20th century equivalent of the heresy trial and that Oppenheimer had been chosen to exemplify the dangers of the scientific heretic to the State. He pointed out that the scientific community was deeply stirred.

And so it was. *The Bulletin of the Atomic Scientists* had had little difficulty in recruiting for its columns sheets of testimony to Oppenheimer's integrity and loyalty from America's most eminent scientists. Some who did not contribute could not do so because they were to appear at the hearings. Bethe, Bush, Conant, Fermi, General Groves and many of the others who, during the war years and after, had worked at close quarters with Oppenheimer, turned up as witnesses in Washington and paraded themselves before the Board to sing his praises. But never at any point was the weight of opinion such that it seemed that there was no case to answer. That stacked on the other side of the balance was also heavy.

One of the incidents in Oppenheimer's story which rested heavily against him did so not because of the intrinsic value of the evidence it brought against him, but because of the flaw in character which it revealed. At first it had the appearance of a red-herring; in the end it contributed strongly to the case against him.

It was known as the Chevalier incident. It critically affected the lives of both men directly involved in it. It occurred one evening in the winter of 1942–3 when Oppenheimer and his wife invited to dinner at their house on Eagle Hill in Berkeley a professor of French at the University and his wife.

The teacher was Haakon Chevalier. During the four or five years during which he and Oppenheimer had known each other, Chevalier's relationship with the physicist had grown from a distant respect based on the earliest legends and myths already being built round the man into an uncritical admiration. For them both the friendship had been based on common interests in the left-wing academic west coast political activities; but for Chevalier at least the bonds had gone deeper. Oppenheimer's singular character, his outstanding intelligence and the ability which went along with it, were powerful charms as they were to many others before and since. Chevalier was flattered to find the friendship returned, even though he knew he could never hope to collect it in the boundless fashion it was given. At a later time, when to all intents and purposes the friendship had died, he was to write to Oppenheimer: 'I have loved you as I have loved no other man. I placed in you an absolute trust. I would have defended you to the death against malice or slander.'[4]

That evening, as dinner was being prepared, Oppenheimer went into the kitchen to mix the drinks. Chevalier followed; he wished to get away from the wives and talk to Oppenheimer alone about a conversation he had had with an Englishman named George Eltenton. Eltenton, a chemical engineer who had spent several years working in Russia for a British firm, was now employed by the Shell Development Company working and living in the United States. He knew, as did Chevalier, that Oppenheimer was engaged on secret scientific war work, and suggested to Chevalier that Oppenheimer, also known to Eltenton for his left-wing sympathies, might be in favour of closer scientific co-operation with Soviet scientists and might even wish to promote collaboration. He suggested that Chevalier should sound out Oppenheimer on the subject.

Chevalier reported the conversation and has now little recall of how Oppenheimer responded other than to appear obviously disturbed by its allusions. Oppenheimer's recollection of Chevalier's message was,

. . . That Eltenton had told him that he had a method, he had means of getting technical information to Soviet scientists. He didn't describe the means. I thought I said, 'But that is treason', but I am not sure. I said anyway something, 'This is a terrible thing to do.' Chevalier said or expressed complete agreement. That was the end of it. It was a very brief conversation.[5]

They left the kitchen carrying drinks for their wives, the incident, it seemed, finished.

Had Oppenheimer heard these sentences spoken a few months earlier in the house on Eagle Hill, in all probability he would have gone out of the kitchen and quickly lost them from his consciousness. But it was only because he was now on the verge of beginning a new life away from Berkeley that they had been spoken at all, and in this new life these sentences could be given the most sinister of interpretations; they stayed fixed in his mind.

Oppenheimer's acceptance of his role of scientific leader of Los Alamos meant that he also must accept the restrictions of personal liberties which scientists and military alike on the mesa found imposed; but he also had to accept that, as key-man, more than most, his private life would be under 24-hour scrutiny by military security. And the scrutiny began even before he left Berkeley for the New Mexican laboratory. At the hearings Oppenheimer had to hear accounts of his visit to his ex-fiancée, Jean Tatlock, which had been well noted by security men (just as had the fact that she was a communist), and that he had spent the night with her before leaving for Los Alamos.

The left-wing background of each member of his family and of his close friends was examined as closely as was his own. His letters were opened, his conversations recorded and, as one security man put it, 'all sorts of nasty things' done to ensure the security and the loyalty of the man at the top of the Manhattan Project.

The investigators were soon disturbed by what they dug up about the left-wing connections of Robert Oppenheimer. In detail they had checked any leads he had chosen to give, and had paid the closest attention to his routine security interviews. When, in August 1943, a junior officer reported to Lt Colonel Borish Pash, Chief of Counter-intelligence, that Oppenheimer had casually mentioned some months-old important information concerning espionage, hackles in the security department were already halfway risen. Pash immediately arranged to interview the physicist.

The information Oppenheimer had to give derived from the Chevalier incident.

The conversation between Pash and Oppenheimer was secretly monitored and recorded; played back at the hearing it showed each man unsure of the ground he had chosen to tread, and unsure of the interpretations which the other might put on what he had to say. In essence the story Oppenheimer had to tell Pash was, that a man who had spent a number of years in the Soviet Union – and he named George Eltenton as this man – had attempted to approach *three* members of the Manhattan Project through an intermediary to persuade them to transmit information to the Soviet consul. Repeatedly, Pash tried to get Oppenheimer to identify the project members or the intermediary, as Oppenheimer must have known he would. Repeatedly Oppenheimer refused: the purpose of this tale being (as he later confessed at the hearing) only that of drawing attention to Eltenton's activities without harming whom he believed to be the innocent contacts.

Three weeks later Colonel John Lansdale, who had been appointed by General Groves to be responsible for the overall security and intelligence of the atomic bomb project, followed up the long interview Pash had given Oppenheimer with one of his own. This too was secretly monitored and recorded. Again interrogator and interrogated went through the uncomfortable preliminairies of thrust and parry, Lansdale using the sycophantic techniques of his trade in order to have Oppenheimer name names. He began,

> Well, now I want to say this – and without intent of flattery or complimenting or anything else, that you're probably the most intelligent man I ever met, and I'm not sold on myself that I kid you sometimes, see? And I'll admit freely that at the time we had our discussion at Los Alamos I was not perfectly frank with you. My reasons for not being are immaterial now. Since your discussions with Colonel Pash I think that the only sensible thing is to be as frank with you as I can. I'm not going to mention certain names, but I think that you can give us an enormous amount of help, and as I talk you will realise, I think, some of the difficulties that have beset us.[6]

But the soft-sell of the preliminaries had little effect when Lansdale moved on to the main purpose of the meeting: the identity of the intermediary.

Lansdale: Well now, you see what you stated that he contacted, I believe it was three persons on the project, and they told him to go to hell in substance.

Oppenheimer: Although probably more politely.

Lansdale: And how do you know that he hasn't contacted others?

Oppenheimer: I don't. I can't know that. It would seem obvious that he would have.

Lansdale: If you heard about them they unquestionably were not successful.

Oppenheimer: Yes.

Lansdale: If you didn't hear about them they might be successful or they might at least be thinking about it, don't you see? Now you can, therefore, see from our point of view the importance of knowing what their channel is.

Oppenheimer: Yes.

Lansdale: I was wondering, is this man a friend of yours by any chance?

Oppenheimer: He's an acquaintance of mine, I've known over many years.

Lansdale: Well do you – I mean there are acquaintances and there are friends. In other words, do you hesitate for fear of implicating a friend?

Oppenheimer: I hesitate to mention any more names because of the fact that the other names I have do not seem to be people who were guilty of anything or people who I would like to get mixed up in it, and in my own views I know that this is a view which you are in a position to doubt. They are not people who are going to get tied up in it in any other way. That is, I have a feeling that this is an extremely erratic and unsystematic thing.

Lansdale: Here is, I want you to in no derogatory way understand my position again.

Oppenheimer: Well . . . there is a very strong feeling. Putting my finger on it I did it because of a sense of duty. I feel justified. . . .

Lansdale: Now, here is an instance in which there is an actual attempt of espionage against probably the most important thing we're doing. You tell us about it 3 months later.

Oppenheimer: More than that, I think.

Lansdale: More than that. When the trail is cold it's stopped, when you have no reason not to suppose that these cases which you hear about are unsuccessful, that another attempt was made in which you didn't hear about because it was successful.

Oppenheimer: Possibly. I am very, very inclined to doubt that it would have gone through this channel.

Lansdale: Why?

Oppenheimer: Because I had the feeling that this was a cocktail party channel A couple of guys who saw each other more or less by accident.

Lansdale: Well, people don't usually do things like that at cocktail parties. . . .[7]

Lansdale could wean no name connected with the incident other than that of Eltenton. It was only when General Groves himself later confronted Oppenheimer, and to all intents ordered him to name the contacts and intermediaries, that Oppenheimer told what he claimed to be the true story: that the only contact had been attempted through him, Oppenheimer, with Chevalier acting as intermediary.

Throughout the Hearing Roger Robb, the Board's counsel, relentlessly returned to this incident and the embroidered version of it which Oppenheimer had given the security men.

> *Robb:* Why did you go into such great circumstantial detail about this thing if you were telling a cock and bull story?
> *Oppenheimer:* I fear that this whole thing is a piece of idiocy. I am afraid I can't explain why there was a consul, why there was microfilm, why there were three people on the project, why two of them were at Los Alamos. All of them seems wholly false to me.
> *Robb:* You will agree, would you not, sir, that if the story you told to Colonel Pash was true, it made things look very bad for Mr Chevalier?
> *Oppenheimer:* For anyone involved in it, yes, sir.
> *Robb:* Including you?
> *Oppenheimer:* Right.
> *Robb:* Isn't it a fair statement today, Dr Oppenheimer, that according to your testimony now you told not one lie to Colonel Pash, but a whole fabrication and tissue of lies?
> *Oppenheimer:* Right.
> *Robb:* In great circumstantial detail, is that correct?
> *Oppenheimer:* Right.[8]

The Board was much perturbed by the evidence it had heard and the distressing light this had thrown on Oppenheimer's character. Chairman Gray attempted to summarise what he and his two colleagues understood of the affair: according to Oppenheimer, Chevalier was his good friend whom he believed to be innocent of sinister intent and whom he had tried (unsuccessfully as it turned out) to protect from the harassment of an investigation by security, by withholding his name as the intermediary – as well as withholding his own name as the contact. The cock and bull story was fabricated, Oppenheimer claimed, in order to disclose the name of Eltenton so that, if Eltenton were involved in espionage, security could investigate him as it wished.

Gray now asked of Oppenheimer, if Chevalier was his friend, why then did he invent a complicated false tale which might show

U

Chevalier, along with others, to be even more deeply implicated in a conspiracy? Was it possible that the story he had told Pash and Lansdale was true, and the current version he was offering to the Board false?

To the second question Oppenheimer answered that he was now telling the truth. To the first he attempted no rationalisation; he had said what he had said because he had made a culpably bad human error of judgment: in his own words, because he had been an idiot.

But the idiocy of Oppenheimer's behaviour was not the Board's chief concern with the Chevalier incident. Its concern was now that Oppenheimer was a self-confessed liar. If necessary the evidence connected with the affair could have been followed up in more detail to attempt to decide which of the two stories concerning Chevalier was true, and what harm had come of which lie. Even at the time of the hearings security investigations of Chevalier and Eltenton had led to no prosecution being made against either, leaving little doubt that, as an isolated incident, the business was no more than a storm in a teacup.

But the Chevalier affair did not stand alone. Ranged along with it was a great mound of conflicting evidence concerning Oppenheimer's left-wing leanings and his attitude to the thermonuclear weapon. What was to be believed about the rest of the evidence he was offering during his long days before the Board?

Whatever other explanations Oppenheimer gave for his attitudes, he never chose to offer to his judges a false suffering face of troubled morality. He took some care to explain that even in 1942 he had understood that he was engaged in what he described as 'a potentially world-shattering undertaking': the groundwork which would lead to Alamogordo and Eniwetok. Even during that same week when Hiroshima and Nagasaki were being bombed, he pointed out, he had been involved in discussions at Los Alamos when atomic war heads for guided missiles, improvements in bomb designs as well as the thermonuclear programme were being sketched out for the future. And though he detached himself from bomb-making in 1945, when in 1947 the future of the Los Alamos laboratory was undergoing reappraisal, he lent his support to its main purpose; morale, on this occasion, rather than morals, was his concern.

We suggested that every inducement be made available to make work at Los Alamos attractive in the way of salaries and housing, but above all in the morale sphere in the sense of giving the men who were there the feeling that they were doing something vital for their country and in getting abroad in the country the sense that Los Alamos was not something left over from the last war, that work on the atomic bomb was somehow not an entirely creditable occupation, but quite the contrary feeling that there was nothing the nation needed more.[9]

On the subject of the hydrogen bomb, in answer to Robb's insistent prodding, he denied ever opposing the project after Truman's decision to proceed with it as a matter of national policy. And in reply to Chairman Gray's wish for an unequivocal statement on whether he attempted to persuade anyone not to work on the thermonuclear weapon, he gave an unqualified denial.

But in spite of the fact that a glittering list of American nuclear physicists presented themselves to the Gray Board to sing Oppenheimer's praises, to endorse his character, to detail his great abilities, to testify to his loyalty and to confirm his own statements on his attitude to the hydrogen bomb, there also appeared a small but critical minority of scientists with some serious reservations scattered through its evidence. Included in this minority was the man who, just as in popular terms Oppenheimer had been called 'the father of the atom bomb', was himself known as 'the father of the hydrogen bomb': Edward Teller.

Before he took the oath at the hearings, the Board had already heard of the part which this exceptional physicist had played in the pro-hydrogen bomb lobby. It had also heard passing references to some of the brushes which the individualist Teller had had with some of his colleagues during his career. Gordon Dean had spoken of him as a genius, and as was often the case with geniuses, a very difficult man to work with. Hans Bethe had described how Teller's inability to accept the agreed line of wartime work had made it necessary to allow him to follow up his own ideas with his own group outside Bethe's main Los Alamos theoretical physics division. Norris Bradbury had described his post-war personal differences of opinion with Teller and disagreements over the administration of the thermonuclear programme. Teller himself, when his turn came to testify, admitted to disagreements with Oppenheimer at Los Alamos adding, 'I believe that it is quite possible, probably, that this was my fault'. He nonetheless paid tribute to Oppenheimer's outstanding leadership of the wartime laboratory.

And when faced with the inevitable question from Roger Robb concerning Oppenheimer's loyalty, Teller joined with the large majority of his physicist colleagues in affirming it. However, in answer to Robb's follow-up of, 'Do you or do you not believe that Dr Oppenheimer is a security risk?', his answer was clear and damning:

> In a great number of cases I have seen Dr Oppenheimer act – I understood that Dr Oppenheimer acted – in a way which for me was exceedingly hard to understand. I thoroughly disagreed with him in numerous issues and his actions frankly appeared to me confused and complicated. To this extent I feel that I would like to see the vital interests of this country in hands which I understand better, and therefore trust more.
> In this very limited sense I would like to express a feeling that I would feel personally more secure if public matters would rest in other hands.[10]

He went on to tell how, soon after the explosion of the first Russian atomic bomb, he telephoned Oppenheimer to ask for advice, and how the curt response had been for him to 'keep your shirt on'; of how, shortly after the presidential decision on the thermonuclear weapon, he asked Oppenheimer for names of physicists who might be willing to be recruited to the project and how these men, all of them at Oppenheimer's Princeton Institute, rejected Teller's advances; of how, as far as he knew, even after January 1950, Oppenheimer was neither supporting nor approving the thermonuclear programme; and of how he believed, if in 1945, people such as Dr Oppenheimer had lent moral support to Los Alamos and the new programme, then the hydrogen bomb would have been built four years earlier than in fact it was.

When Teller had completed his evidence Gordon Gray put to him the question which he and his two colleagues had been appointed to answer: did Teller believe that Oppenheimer's security clearance should be granted? Teller had in effect already answered, but he repeated himself:

> I believe, and that is merely a question of belief and there is no expertness, no real information behind it, that Dr Oppenheimer's character is such that he would not knowingly endanger the safety

of this country. To the extent, therefore, that your question is directed toward intent, I would say I do not see any reason to deny clearance.

If it is a question of wisdom and judgement, as demonstrated by actions since 1945, then I would say one would be wiser not to grant clearance. . . .[11]

From the Board Dr Evans had two final questions to ask of Teller.

> *Evans:* You understand, of course, that we did not seek the job on this board, do you not?
> *Teller:* You understand, sir, that I did not want to be at this end of the table either.
> *Evans:* I want to ask you one question. Do you think the action of a committee like this, no matter what it may be, will be the source of great discussion in the National Academy and among scientific men in general?
> *Teller:* It already is and it certainly will be.
> *Evans:* That is all I wanted to say.[12]

Teller was not the only physicist of standing to interpret Oppenheimer's attitude to the thermonuclear weapon as being against the best interests of his country. Luis Alvarez, like Teller, had known Oppenheimer well since the ideas of super-bombs were first being mooted, and it was at that time during the war years the Los Alamos leader had become what Alvarez called one of his 'scientific heroes'. It was Oppenheimer who in 1942 had first described to him the possibility of a thermonuclear weapon, and the limitless magnitude which might be expected from its explosion. But after the first Russian atomic bomb, along with Teller and Ernest Lawrence, Alvarez had developed considerable dissatisfaction with the inadequate status of the thermonuclear programme. When during the hearings Alvarez's turn came to give evidence he left no doubt in the minds of the Board that in his opinion Oppenheimer's influence opposing the hydrogen bomb had been powerful and that those of his scientific colleagues who had agreed with his one-time hero had borne the recognisable stamp of this influence.

During the period 1951 and 1952 when optimism among the proponents of the thermonuclear weapon was beginning to ride high, David Griggs had been appointed chief scientist of the Air Force. Whilst he had held this office he had learnt of security's suspicions of Oppenheimer's loyalty. At the hearings he went on record as saying 'Dr Oppenheimer is the only one of my scientific acquaintances about whom I have ever felt there was a serious question as to their loyalty.'

There were others who came forward to add their testimony to the evidence mounting against Oppenheimer. Equally, there were those who were prepared to refute this testimony in no uncertain terms, and in so doing demonstrate to the Gray Board and to the outside world the depth of feeling and personal bitterness which the issues now being exposed had so painfully precipitated into the scientific community. The M.I.T. physicist, Jerrold Zacharias, on being recalled to give evidence after Griggs's appearance was prepared to tell the Board that his respect for Griggs had declined rapidly over the past two years and that Griggs's sworn testimony had taken his respect to a new low point. Vannevar Bush too put in a second appearance to refute some of the evidence Alvarez had offered on Oppenheimer. Scientist was ranged again scientist and it was inevitable that no matter what the ultimate decision of the Board would be in the matter of Robert Oppenheimer, he would not be the only man to carry away scars from this experience.

Throughout the proceedings, member of the Board Ward Evans showed acute awareness of the unpleasant consequences which the hearings would be bound to have on the scientific community and on its relationships with the world at large. Frequently he interrupted the tense, protracted daily routine of the inquiry with interjections which he called 'nothing particularly pertinent to this proceeding': questions about old mutual friends, whether they were in good health, and whether the witness would pass on his regards. They were questions designed to remind those present, and perhaps to reaffirm to his colleagues on the Board, that the affair was a hearing and not a trial, and to humanise the legal-scientific clashes which each day produced.

As the only scientist member of the Board he took on himself the task of eliciting for his colleagues' benefit, and also for the benefit of some of those involved in the proceedings who apparently carried doubts, the fact that scientists were a part of the common culture with abilities and qualities similar to the rest of mankind. He was loath to let assumptions to the contrary pass by unquestioned. On one occasion the Los Alamos security officer, Colonel Lansdale, had said:

The scientists en masse presented an extremely difficult problem. The reason for it, as near as I can judge, is that with certain outstanding exceptions they lacked what I called breadth. They were extremely competent in their field but their extreme competence in their chosen field led them falsely to believe that they were as competent in any other field.

The result when you got them together was to make administration pretty difficult because each one thought that he could administer the administrative aspects of the Army post better than any Army officer, for example, and didn't hesitate to say so with respect to any detail of living or detail of security or anything else. I hope my scientist friends will forgive me, but the very nature of them made things pretty difficult.[13]

If Evans forgave he did not forget and later steered Lansdale back to the subject.

Evans: Do you as a rule dislike the scientific mind? Is it a peculiar thing?
Lansdale: I will say this, that during the war I came very strongly to dislike the characteristics which it exhibited.[14]

The professor of chemistry returned to the theme again with the physicist Norris Bradbury.

Evans: Do you think that scientific men as a rule are rather peculiar individuals?
Bradbury: When did I stop beating my wife?
Gray: Especially chemistry professors?
Evans: No, physics professors.
Bradbury: Scientists are human beings. . . . I think you are likely to find among people who have imaginative minds in the scientific field, individuals who are also willing, eager to look at a number of other fields with the same type of interest, willingness to examine, to be convinced and without *a priori* convictions as to rightness or wrongness, that this constant or this or that curve or this or that function is fatal.

I think the same sort of willingness to explore other areas of human activity is probably characteristic. If this makes them peculiar, I think it is probably a desirable peculiarity.[15]

How then should a scientist in Oppenheimer's perilous position be judged? Should he be treated more stringently than his fellow men because his ability in a particular scientific field made him peculiarly more responsible to society for his actions? Or should he be accorded more compassion because he had freely chosen to take on such grave responsibilities in contributing to the construction of weapons of war on society's behalf?

Lansdale, the non-scientist, considered scientists to be narrow beings with consequently undesirable characteristics. Bradbury, the scientist, drew precisely the opposite conclusion. Perhaps as a result of forcing his interchanges with these men Evans could claim to have demonstrated that the truth lay between these extremes: that scientists are ordinary mortals, and should be judged as such and not according some special code especially laid down for them; undoubtedly some scientists had exceptional abilities in some fields, and undoubtedly some had failings in others; the non-scientists might well look on and say, 'There, but for the grace of God, go I'.

The Gray Board heard its evidence at a time when attitudes towards a man of Oppenheimer's background were strongly influenced by the climate of the times. In 1954 the atmosphere which McCarthyism had bred was such that any man tainted with communist influence was suspect. How would evidence for and against Oppenheimer have been received at any other time in post-war history? It could be argued that much of the derogatory information concerning Oppenheimer's left-wing past was known when he was last cleared for security in 1947. The Board was now judging him on the basis of this same information. Why should it now reach a different conclusion? By the same token, the evidence of the Chevalier incident, which had been used to suggest that Oppenheimer, the head of the Los Alamos scientific project, was involved in espionage, was in 1954 considered by many to be irrelevant and even laughable. Yet this might not have appeared so ludicrous in 1967, the year Kim Philby revealed himself to be an officer of the Russian K.G.B.

The Board completed its hearings on 6th May 1954 after recording three-quarters of a million words of evidence. Its findings began: 'It must be understood that in our world in which the survival of free institutions and of individual rights is at stake, every person must in his own way be a guardian of the national security.'[16]

After considering individually each of the allegations of Nichols original letter to Oppenheimer it summarised its conclusions:

> In arriving at our recommendation we have sought to address
> ourselves to the whole question before us and not to consider the
> problem as a fragmented one either in terms of specific criteria or

in terms of any period in Dr Oppenheimer's life, or to consider loyalty, character, and associations separately.

However, of course, the most serious finding which this Board could make as a result of these proceedings would be that of disloyalty on the part of Dr Oppenheimer to his country. For that reason, we have given particular attention to the question of his loyalty, and we have come to a clear conclusion, which should be reassuring to the people of this country, that he is a loyal citizen. If this were the only consideration, therefore, we would recommend that the reinstatement of his clearance would not be a danger to the common defense and security.

We have, however, been unable to arrive at the conclusion that it would be clearly consistent with the security interests of the United States to reinstate Dr Oppenheimer's clearance and, therefore, do not so recommend.

The following considerations have been controlling in leading us to our conclusion:

1. We find that Dr Oppenheimer's continuing conduct and associations have reflected a serious disregard for the requirements of the security system.

2. We have found a susceptibility to influence which could have serious implications for the security interests of the country.

3. We find his conduct in the hydrogen-bomb program sufficiently disturbing as to raise a doubt as to whether his future participation, if characterised by the same attitudes in a Government program relating to the national defense, would be clearly consistent with the best interests of security.

4. We have regretfully concluded that Dr Oppenheimer has been less than candid in several instances in his testimony before this Board.[17]

In its general considerations leading to its recommendation the Board had said in regard to Oppenheimer's part in the thermonuclear programme:

> We cannot dismiss the matter of Dr Oppenheimer's relationship to the development of the hydrogen bomb simply with the finding that his conduct was not motivated by disloyalty, because it is our conclusion that, whatever the motivation, the security interests of the United States were affected.
>
> We believe that, had Dr Oppenheimer given his enthusiastic support for the program a concerted effort would have been initiated at an earlier date. Following the President's decision he did not show the enthusiastic support for the program which might have been expected of the chief atomic adviser to the Government under the circumstances. Indeed, a failure to communicate an abandonment of his earlier position undoubtedly had an effect upon other scientists. It is our feeling that Dr Oppenheimer's

influence in the atomic scientific circles with respect to the hydrogen bomb was far greater than he would have led this Board to believe in his testimony before the Board.[18]

The report was signed by Gordon Gray and Thomas Morgan. The third member of the Board, Ward Evans, submitted his own minority findings. He was at serious odds with his non-scientist colleagues.

Of Oppenheimer he said:

> His judgment was bad in some cases, and most excellent in others but, in my estimation, it is better now than it was in 1947 and to damn him now and ruin his career and his service, I cannot do it. His statements in cross examination show him to be still naïve, but extremely honest and such statements work to his benefit in my estimation. All people are somewhat of a security risk. I don't think we have to go out of our way to point out how this man might be a security risk.[19]

He went on to state his belief that Oppenheimer had not hindered the development of the H-bomb and that there had been absolutely nothing in the testimony to show that he had. His personal opinion was,

> . . . that our failure to clear Dr Oppenheimer will be a black mark on the escutcheon of our country. His witnesses are a considerable segment of the scientific backbone of our Nation and they endorse him. I am worried about the effect an improper decision may have on the scientific development in our country.[20]

He suggested that Dr Oppenheimer's clearance be restored.

On 28th June the Atomic Energy Commission took its vote on the recommendation of the Gray Board. By a majority of 4 to 1 it elected to deny security clearance. The dissenting vote was again that of a scientist, Henry Smyth. On the question of whether Oppenheimer's opinions had affected policy in the 1949 governmental debate on whether the United States should manufacture a hydrogen bomb, it was stated that: 'Neither in the deliberation by the full Commission nor in the review of the Gray Board was importance attached to the opinions of Dr Oppenheimer . . .' for 'In this debate, Dr Oppenheimer was, of course, entitled to his opinion.' Rather, he was considered not to be entitled to 'the continued confidence of the Government and this Commission because of the proof of fundamental defects in his "character" '.

It was this same singular character of the man which had been so influential in the 10 years of his governmental service; it had inveigled some into passionate admiration, others into chary suspicion; it had dominated one of the greatest projects in which science had been applied to war, just as it had dominated the concern of those present in the quasi-courtroom of the Gray Board. What it had achieved in the past must now be overlooked, for in 1954 it was found to be wanting.

The matter of J. Robert Oppenheimer was ostensibly closed: but not so the issues involved. By its recommendation the Gray Board had concluded that scientists involving themselves in matters affecting the security of the State bore special responsibilities. Such is the case of course with all intellectuals who, for whatever reason, involve themselves in governmental business. But this was not all in the case of Robert Oppenheimer. By its findings on the hydrogen bomb, the Board's majority decision was that a scientist, by virtue of his esoteric knowledge of a weapon of war, bore responsibilities far beyond those of any of his fellow citizens. Oppenheimer was found to be loyal to his country. Indeed, there was an enormous debt owing to him by the United States for the part he had played in bringing the atomic bomb to fruition. But in spite of this loyalty, because his opinion in the case of the thermonuclear weapon so strongly affected the attitude of other scientists and so slowed its initiation, and because the Gray Board considered the advice he had given in respect of it to be bad advice, no matter with what good faith he had intended, he was considered to be culpable. The Gray Board had concluded that in this case this scientist should be judged according to a special code. Oppenheimer had committed heresy. He should pay the price.

As Ward Evans predicted, the effect on science of this affair was deeply disturbing. In their considerations his two colleagues had claimed to have been heartened by the manner in which many scientists had sprung to the defence of Oppenheimer. Nevertheless, they felt constrained to express their concern that 'in this solidarity there have been attitudes so uncompromising in support of science in general, and Dr Oppenheimer in particular, that some witnesses have, in our judgement, allowed their convictions to supersede what might reasonably have been their recollections'.[21]

Initially it seemed that the result of the hearings was a great defeat for the scientific majority; one of its respected, and at times revered, leaders had had his career laid waste, his best motives challenged and his actions denigrated. He had been brought to heel: the rest of the pack could take warning. But for all that he suffered, Oppenheimer did not pillory himself in vain. Details of the Gray Board's three weeks of microscopic investigation of the life, times and shortcomings of this scientist, received world-wide publicity. For the first time the difficulties of the role of the scientist in wartime had been brought prominently into the public domain: the difficulties of men as creators of weapons, and as advisers of how society should best use these weapons. Oppenheimer had recognised that his duties did not end with his contributions to the creative act. The demands of society on its scientists had so changed throughout the century, that society had not expected his responsibilities to end there. The advice he had given had been flawed and he had suffered a loud public rebuttal. So loud had it been that many listened who might not before have heard of the existence of the scientist's dilemma. Oppenheimer, in defeat, had ineradicably emphasised the importance of his breed; its deep foot-print had been with the 20th century from its very beginnings, but only now could many 20th-century common-men see it leading them on. If it did not show signs of being sure-footed then the future of the second half of the century could well be in doubt, such was the burden that men such as Oppenheimer have had, and will have, to bear. Was it right to dismiss so summarily from the scene a man capable of so many more guiding steps?

From science there had been uncompromising attitudes on both sides and the scars gouged by this affair were to remain tender for some time to come. Shortly after the Hearing George Gamow visited Edward Teller at Los Alamos and the two physicists sat talking at the window of Teller's dining room. Walking past in the street outside there appeared two of the laboratory's leading physicists who waved to Gamow. Teller rushed out to the door to try to speak to them, but they continued to walk on as though they had not heard.

The personal animosities which the Gray Board had brought to the surface, and in some cases created, worried many of those who believed most strongly in the values of a united scientific community. It was reported that shortly before his death from cancer in November 1954 Enrico Fermi, from his sick-bed, attempted to persuade Teller to try to mend the great rift which had hived off one section from the other.

Time itself is a great healer, and as the years passed, and as the climates of both political and scientific opinion changed, the wounds began to take on a less bloody appearance. But though Oppenheimer accepted the result of the hearings quietly and with dignity, those who knew him intimately found its consequences had brought on a permanent change. The vitality of his brilliant mind was quenched and his active work in physics diminished. Many recognised that there still remained some mend to be made and that the United States had a great debt still to discharge to this man. It was in partial acknowledgement of this that President Kennedy, on the day he died, announced his decision to award the Enrico Fermi Medal of the Atomic Energy Commission, and its 50,000 dollar prize, to Robert Oppenheimer. Edward Teller, who had himself received the award from Kennedy in 1962, had added his name to the ballot paper nominating the man his evidence had helped condemn.

'I think it is just possible, Mr President,' Oppenheimer said to Lyndon Johnson who took over the task which Kennedy had set aside for himself, 'that it has taken some charity and some courage for you to make this award today. That would seem to me a good augur for all our futures.' But when Oppenheimer died on 18th February 1967 his government had found no excuse to return his security clearance to him.

15 : The Shape of War to Come

In answer to a question from Dr Ward Evans at the Oppenheimer hearings, Luis Alvarez, the physicist, said:

> I have never had any moral scruples about having worked on the atomic bomb, because I felt that the atomic bomb saved countless lives, both Japanese and American. Had the war gone on for another week, I am sure that the fire raids on the Japanese cities would have killed more people than were killed in the atomic bombs. . . .
>
> *Evans:* Don't we always have moral scruples when a new weapon is produced?
>
> *Alvarez:* That is a question I can't answer, sir.
>
> *Evans:* After the Battle of Hastings, a little before my time . . . there was great talk about ostracising the long bow, because it was so strong that it could fire an arrow with such force, it occasionally pierced armor and killed a man. They felt they ought to outlaw it. When the Kentucky rifle came in, it was so deadly that they talked of getting rid of it. When we had poison gas, I made a lot of lectures about it, that it was terrible. So we have had that after every new weapon that has been developed.[1]

Evans's knowledge of the weapons of 1066 and all that was suspect; but his point that each new weapon to alter the nature of warfare is inevitably denounced as being more inhuman than any to precede it is valid. The weapon which in his youth had been denounced for the cruelty and the suffering it caused was poison gas. The attitudes of horror and revulsion which appeared after its first large-scale use in 1915, and which he shared, still persist today.

In 1943, during the next war President Roosevelt stated his

own nation's position with respect to this type of weapon. It was based on these attitudes. He said:

> From time to time since the present war began there have been reports that one or more of the Axis Powers were seriously contemplating use of poisonous or noxious gases or other inhumane devices of warfare.
>
> I have been loath to believe that any nation, even present enemies, could or would be willing to loose upon mankind such terrible and inhuman weapons. However, evidence that the Axis Powers are making significant preparations indicative of such an intention is being reported with increasing frequency from a variety of sources. Use of such weapons has been outlawed by the general opinion of civilised mankind. This country has not used them, and I hope that we never will be compelled to use them. I state categorically that we shall under no circumstances resort to the use of such weapons unless they are first used by our enemies.[2]

But not all of Roosevelt's countrymen were so resolutely opposed to anything other than retaliatory use of these weapons, nor did they acknowledge any clear division between 'humane' and 'inhumane devices of war' which their President had apparently seen fit to define. Several American Generals were appalled at the losses suffered in the last year of the Pacific War. In the battle of Iwo Jima there were 27,000 U.S. casualties, including 6,775 dead, and practically all the 21,000 Japanese defenders of the island were killed or committed suicide. In the weeks which followed, but still two weeks before Roosevelt's death in April 1945, General Marshall had been prepared to use gas at Okinawa in order to prevent the likely repetition of the slaughter of his own troops. David Lilienthal recorded in his Journal how, after the war, Marshall had admitted to him that the reason it was not used had been

> '. . . chiefly the strong opposition of Churchill and the British. They were afraid that this would be the signal for the Germans to use gas against England.' Then he spent some time indicating how exceedingly vulnerable the British are, compared to ourselves and others; how well they realise it; how they worry about it; how it affects their attitude in international dealings; how important it is for us to understand their feeling.[3]

Marshall apparently made no mention of the fact that Britain, along with a number of other countries, had signed and ratified the Geneva Protocol of 1935, which prohibited chemical and bacteriological weapons, but that the United States, which had signed the protocol, had failed to ratify it. Roosevelt's statement of 1943 was all the Axis Powers had to rely on that the United States would not use the weapons his words clearly showed she possessed.

The United States Army's definitions of chemical and biological warfare – or CBW, as the innocuous-sounding mid-20th century abbreviation has it – are the following:

> *Chemical Warfare (CW) – Tactics and technique of conducting warfare by use of toxic chemical agents.* This includes the use of gas, smokes, flame and incendiaries.
> *Biological warfare (BW) – Employment of living organisms, toxic biological products, and chemical plant growth regulators to produce death or casualties in man, animals, or plants.*

Although during the first quarter of the century the art of the chemist, and during the second quarter that of the biologist, developed sufficiently to enable him to prepare highly sophisticated and frightening weapons which fit into these categories, fundamentally there is nothing new or particular to this century in their use in warfare. Unlike the technologies of the long-range rocket or the nuclear bomb those of chemical and biological warfare have a long rooted past. Poison gas was sustaining man's inhumanity to man long before the birth of Christ: during the Peloponesian War the suffocating fumes from burning pitch and sulphur put paid to many a Spartan and Athenian. Incendiaries such as 'Greek fire', probably a mixture of combustible chemicals with a light distillate of oil, were in use for several centuries up to the middle ages. Charles XII of Sweden's troops used the smoke of damp, burning straw to cover their movements in 1700. And gunpowder, itself a chemical explosive, was used mixed with arsenic shortly after its earliest development by some primitive Chinese chemist in the 9th century A.D.

There is an equally long, colourful and lethal line in biological weaponry. It was believed that the Spartans ravaged the Athenians with typhoid by spoiling supplies of drinking water. During the siege of Mantua, Napoleon flooded low-lying land in the hope of spreading malaria among the Italian citizenry. And in the American Civil War it was commonplace for retreating troops to drive animals into ponds and streams and shoot them there so that the putrescent flesh would poison the ascendent enemy's water.

Thus well before the coining of the word 'scientist' in the early 19th century, men with a bent towards certain aspects of natural

philosophy were applying their specialised knowledge to decimat-
ing their fellow creatures by CBW But only in time for the Great
War, when the 'scientist' had emerged with the industrial society,
were techniques and abilities sufficiently refined for chemical warfare to be considered for use on a large scale.

France had been credited as the first nation to use it in that war. In
August of 1914 tear-gas grenades fired from rifles were put down
on the German lines. These contained ethylbromacetate, a compound rated by chemical warfare experts to be at least as toxic as
chlorine. German Artillery retaliated with tear gas artillery shells.

The gas used by the Germans in the now notorious attack on
the Ypres salient on 22nd April 1915 was chlorine itself. It is this
attack which is commonly looked on as the first in the history of
modern chemical warfare. It devastated the French and British
troops who were taken unawares as a choking green-yellow
vapour spread over their trenches in a cloud 6 kilometres long
and 600–900 metres deep. It was discharged from thousands of
steel cylinders in the German lines using a technique devised by
Fritz Haber.

With what to senior German officers were laughably scant
facilities, Haber succeeded in breaking the resistance of a previously impassable French division. But the military leaders, having
little trust in the chemist's experiment, provided him with no
reserve of troops. The gap in the Allied lines was not exploited and
British reserves were able to move in to close it.

Now that the gas war had begun in earnest chemists from
behind both lines were on call to provide new agents. In 1916
phosgene appeared. It was less painful but more lethal than
chlorine. In 1917 mustard gas, the most effective of all Great War
chemical agents, took the field. It penetrated cloth, attacked eyes
and lungs, and blistered the skin. As protection, or part protection, from the one side was produced, the other countered with a
new gas or vapour. Between 1915 and 1918 Germany and Austria
used 66,000 tons of toxic chemicals in their gas-warfare and the
Allies only fractionally less.

With the range of useful weapons now on view, every school-
boy chemist could try his hand at the experimental procedures if
he had a mind. One Military Intelligence officer serving in the
Middle East in 1917, Colonel Richard Meinertzhagen, who had
used enough vicious methods of killing off his country's enemies

to cause even T. E. Lawrence to purse his lips, quickly saw the chances for powerful narcotics. He arranged to drop propaganda leaflets and cigarettes drugged with opium over the Turkish lines the night before an attack. He had tried the cigarettes out on himself and had been charmed by the results: his energy had been sapped, he had abandoned himself to dreams of beauty, and had been unable to act or think.

This type of scheme which Meinertzhagen advocated to General Allenby, and which depended on the use of a drug temporarily to incapacitate, and whose long term effects were uncertain, was still being advocated by military men and scientists half a century later. Allenby rejected the scheme because he considered it to be too close to poisoning. Meinertzhagen (who killed many men with his own hands during his long life) put it into operation in spite of his superior officer. His was 'the principle that anything which saved casualties to our own men was justified'. Fortunately the incident was sufficiently obscure not to provoke retaliations in kind.

The Great War coincided with the rise to a new maturity of the physical and chemical sciences and therefore saw the sweeping applications to the methods of war of the resultant technologies, of which gas-warfare was one. The hey-day of the biological sciences was yet to come and between 1914 and 1918 there was little evidence that biological weapons were employed with much considered intent. In 1917, foreseeing the possibilities, Haber founded a technical division attached to the Prussian War Ministry to develop defensive measures against animal and vegeable parasites; and it was reputed that the Germans inoculated horses to be shipped to the Allies with infectious bacteria which caused anthrax and glanders. Otherwise, the war ended with only the fringes of a potentially large-scale war weapon explored.

When World War I ended an intense odium had been incurred towards the 'new' weapons and although millions of men had been killed or maimed by more conventional, and apparently more acceptable means, the horror of gas-warfare has persisted to the present day. The reasons for this attitude lay in the reaction of the Allies to the Ypres attack of April 1915. Unprepared, without adequate defence and fearful of more chlorine attacks on an even grander scale followed up by adequate reserves of German troops, the Allies replied with its one immediately available counter-measure: propaganda. Sensational reports of the sufferings of French, British and Canadian soldiers, drenched by the green gas and cruelly choking, reached the world's press. Blame for this terrible innovation was successfully planted on Germany's doorstep without her armies having benefited one iota from the incident which caused the odium.

Whichever nation involved in World War II would be first to take up chemical and biological warfare on the grand scale knew that forever after it would have to bear the abhorrence of the rest of the civilised world. Ironically, it might be argued that if, between 1939 and 1945, a similar concerted effort to produce weapons, had been concentrated on biology and biochemistry as had been forced upon the physical scientists, it is more than likely that the dramatic advances in knowledge in these fields which occurred in the late 1950's and early 1960's would have taken place many years earlier. Instead, physics produced the atomic bomb, the United States bore the repugnance its use entailed, and domestic atomic power supply reached the homes of the world 10, 20, or even more years before it would otherwise have done.

All the major nations involved in World War II made some preparations for CBW and many chemical and biological agents were explored. After the war Japan alone was arraigned for having employed biological weapons. In a trial in 1949 at Khabarovsk in Russia 12 Japanese military prisoners appeared at a war crimes' trial accused of having been involved in bacteriological attacks on Chinese and Mongolian soldiers and civilians. It was reported at this trial that installations had been built in Manchuria in 1936 to test suitable weapons and that it had functioned until 1945 when it was hastily destroyed in the face of a threatened Russian advance. Chinese and Russian prisoners were said to have been used as experimental subjects, and Chinese cities were reported to have been infected with plague by aeroplanes dispersing millions of specially bred fleas. After the war it was considered that the Axis powers had achieved nothing in methods of biological warfare that could not have been matched by the laboratories which had been working in the field at the same time in the United States and Britain.

Britain's wartime establishments were on a small scale; the Microbiological Research Establishment at Porton in Wiltshire had a scientific staff of only 15. Results nevertheless were arresting. In one experiment in 1942 the island of Gruinard off the north-west coast of Scotland was sprayed with anthrax bacteria; according to Dr C. E. Gordon Smith, the present director of the Micro-biological Research Establishment, speaking in 1967, the purpose of the experiment was:

> . . . to find out about the feasibility of attack because at that time the defence people in this country were very worried as to whether this was a dangerous form of warfare. The experiment showed that it probably was a dangerous form of warfare, and it has led really to the development of work to defend ourselves against this.[4]

According to Dr Smith, the island is today still almost as infected as the day it was sprayed and is likely to remain in this condition for a hundred years.

With the possible exception of attacks attributed to the Japanese in China between 1937 and 1943, gas-warfare was never introduced to the battlefields of the Second World War: the odium of 1915 had stuck. However, a paper with tragic overtones, which appeared in the *Lancet* of 1st June 1968, demonstrated the extent of Japan's involvement in preparations for chemical warfare. In this publication medical research workers at Hiroshima University showed that 33 factory workers who had been engaged in the manufacture of mustard gas between 1929 and 1945 on an island in the Hiroshima prefecture had died since 1952 as a result of respiratory tumours.

But agents of chemical warfare other than gas were widely used in World War II. Louis F. Fieser was one of 20 university professors who in October 1940 was invited by James Conant to co-operate with the newly formed National Defense Research Committee in work on bombs, fuels, poison gases and chemical problems. During the next few years Fieser and his co-workers were responsible for the application of scientific method to the production of many ingenious devices of war, in particular incendiaries. The 'bat-bomb' was one of these. Its principle was simple and at first sight even ludicrous. Thousands of bats which had been put into hibernation by cooling, and to which were attached small incendiary bombs, would be released from aeroplane bomb bays. When the bats reached warm air they would come out of hibernation, fly into buildings, and there the incendiaries would ignite. With flimsy and highly combustible Japanese houses as targets the animate weapons might in theory devastate whole cities.

Fieser wrote later:

> Imagine, then, a surprise attack on Tokyo in which a succession of bombers would operate at high altitude for about half an hour, say starting at midnight, each delivering a load of bat-bombs equivalent to some 3,700 fires. There would be no explosions or fire bursts to give warning, and the bombers would depart. With the activated mechanisms all set for a four-hour delay, bombs in strategic and not easily detectable locations would start popping all over the city at 4 a.m. An attractive picture? All those working on the project thought so.[5]

Research showed that this scheme was far from being ludicrous and its impracticability was not the reason for its abandonment; the reason suggested by Fieser is that the bats might have been suspected by the Japanese as being germ carriers, so leaving the United States open to accusations of being the initiator of biological warfare.

In 1942 Fieser gave the name *napalm* to another of his group's inventions; this was an incendiary gel which was made from gasoline mixed with *na*phthenate and aluminium *palm*itate. Among its properties were its ability to adhere to the skin, to create extensive burns and to cause carbon monoxide poisoning. It was widely used in different forms in World War II; it was the incendiary which saturated Japanese cities in the closing weeks of the war and which caused more deaths than the two atomic raids on Hiroshima and Nagasaki.

Napalm, however, was only to be regarded with the same public revulsion once reserved for gas-warfare when it was put to use in Vietnam, 20 years after it first saw action. In 1967 the *New York Times* ran a short article on the leader of the wartime napalm team. According to this article, Louis Fieser, when interviewed, considered that the limits of his responsibilities as a scientist ended with the weapon's manufacture:

> I distinguish between developing a munition of some kind and using it. . . . You can't blame the outfit that put out the rifle that killed the President. I'd do it again, if called upon, in defense of my country.

The nation which was better prepared than any other for strategic toxic chemical warfare between 1939 and 1945 was

Germany. Even before the war began one of her chemists, Gerhard Schrader, had synthesised in the laboratories of I.G. Farben an insecticide, Tabun, which was quickly recognised as a lethal nerve gas. Schrader saw to it that it was accorded the military security he knew it deserved. It was 20 times more toxic than phosgene, and a few breaths of this not easily detected gas caused death. By the end of the war Germany had a stockpile of 12,000 tons of Tabun ready for use. It was the first of a series of phosphoro-organic compounds, of which Sarin and Soman are even more efficient examples, which was found to act as a deadly systemic poison. Its method of action is to interfere with the body's nerve impulses. Normally these impulses are generated by the release of the compound acetylcholine at nerve endings, causing depolarisation of the membrane; this activates the next nerve, muscle or secretory gland. After the impulse has passed, the acetylcholine is destroyed by an enzyme called acetylcholinesterase. The presence of a nerve agent, such as Tabun, inhibits the action of this enzyme so that acetylcholine accumulates and the junction continues to discharge violently. In physical terms this produces convulsions, paralysis and in particular respiratory paralysis, leading to death by asphyxia.

The Allies knew nothing of this work and although some progress had been made in Britain towards preparing less effective agents, the armies which invaded Europe in 1944, inexperienced as they were in gas protection, would have been unable to resist Tabun had Hitler chosen to use it. One of the prizes for the invading forces was the Tabun production plant and this fell to the Russians. It is more than likely that the chemists who manned this plant were guided back to Russia along with their considerable knowledge, just as at the same time Wernher von Braun with his rocketry expertise was ushered in comfort and safety to the United States.

By the time of the outbreak of the Korean War the range of lethal chemical and biological agents available to both sides had increased enormously, but there is no record of any gases having been used. American front-line commanders, troubled by the hill-side tunnels of the North Koreans and Chinese, were later admitted to have sought permission to employ chemical agents, even if only tear and vomiting gases, but this was refused. There were accusations, however, which could have had serious repercussions had national tempers not been kept in check, that the United States had resorted to biological warfare in Korea and Manchuria. China invited an International Scientific Commission to investigate. Included in the Commission was Joseph Needham, the Cambridge biochemist who spent the Second World War years in charge of a scientific and technological mission in China. Needham concluded with his Italian, Swedish, Brazilian, French and Russian colleagues, and still holds this same view today, that insects had indeed been used as vectors of disease. The Commission's Report was rejected out of hand by the United States. Brigadier General Rothschild, who was Chemical Officer to the Far East Command during most of the action, has denied that biological agents were used in any way in Korea.

In Vietnam gas is clearly admitted to be used. On 24th March 1965, the United States Secretary of Defence, Robert McNamara, stated that South Vietnamese armed forces had been equipped with what he described as 'riot control agents' and that these had been recently put into action on Vietcong who had hidden among villagers of the Phy Yen province. The chemical agents employed, which he pointed out were used by metropolitan police forces in civil disturbances in many parts of the world, were three organo-chlorine compounds: DM, a pepper-like irritant which causes nausea and vomiting, and incapacitates for half an hour to two hours; CN, a tear-gas and irritant which causes violent coughing and is effective for about three minutes; and CS, a more severe agent than CN, which was developed at Porton in Britain, and which incapacitates for about 15 minutes.

But the Defence Secretary's carefully prepared statement was no cushion for the news. Political and press comment and, not only that from Russia and China, was as condemnatory as that which followed the Ypres gas attacks exactly 50 years before. 'Americans', said the *New Statesman*, 'are treating the hapless inhabitants of Vietnam as a living laboratory in which to test their new weapons.'

If the Vietnam peasantry was indeed part of an organic laboratory then the chemical experiments performed on this unhappy breed were not only limited to trials with non-toxic gas. The flesh-burning jelly, Napalm–B, the production of which was reported to be approaching 50,000,000 lbs a month in 1966 in the United States, has been extensively used in incendiary bombs to fire crops, in flame-throwers to raze villages, and in anti-personnel mines and booby traps, to destroy whatever personnel might blunder into them. More effective in the elimination of the Vietcong's food sources, and the jungle which covers him, has been the spraying from aeroplanes of enormous quantities of powerful chemical defoliants; in the same year these left bare over 100,000 acres of padi fields and 500,000 acres of jungle. The full effects on the ecology of this region of South East Asia may not be known for years to come.

For all practical purposes there is an unbroken line in the use of chemical and biological warfare from the beginnings of recorded history to the present day.

In the Western world the scientific institutions and their scientists who have contributed and still contribute to the increasing sophistication of the armoury are very numerous and most various. It was on 4th January 1946 that the public first learnt the purpose of an establishment called Camp Detrick in Maryland. On that day it was announced that the large complex had been a wartime biological warfare centre. One of the men who had turned up for work there a few months after work had begun in 1943 was Professor Theodor Rosebury, a bacteriologist who has since taken a leading part in drawing the attention of his fellow scientists to the ethical problems of CBW. He relates that during the early months at the establishment,

> . . . there was much quiet but searching discussion among us regarding the place of doctors in such work. Most of us were civilians who chose freely to go and work there. Detrick at that time was the kind of technical post in which civilians, in or out of uniform, made all the important decisions; the professional military kept respectfully out of our way. We resolved the ethical question just as other equally good men resolved the same question at Oak Ridge and Hanford and Chicago and Los Alamos. We were in a crisis that was expected to pass in a limited time, with a return to normal values. At Detrick a certain delicacy concentrated most of the physicians into principally defensive operations—*principally* or *primarily*: the modifiers are needed because military operations can never be exclusively defensive. The point is not extenuating. If extenuation is possible, and I think it is, it depends on the factor of time. We were fighting a fire, and it seemed necessary to risk getting dirty as well as burnt.[6]

Fort Detrick, as it is now called, is today the centre of biological warfare research in the United States. Since the war it has become recognised as a first-rank microbiological laboratory with a status in the eyes of many American scientists equivalent to that of a university 'whose past and present staff members' according to Rosebury 'foregather annually to celebrate auld lang syne'. Some of the research carried out at Detrick is of a conventional microbiological research nature and is published, but the greater part of the work undertaken by its 600 graduates in the 1,300-acre complex has been and remains a closely guarded secret.

The chemical weapons sister establishment of Fort Detrick is the Edgewood Arsenal, Maryland, but these two institutions take up only a fraction of the United States' total research and development budget for CBW of well over 100,000,000 dollars. In addition to industry many academic institutions and their scientists have benefited directly or indirectly from governmental CBW funds. Pennsylvania, Cornell, Johns Hopkins, Stanford and Massachusetts Institute of Technology are only a few of the universities which have accepted contracts or sub-contracts for work which can be related to CBW research; some of this work has been published and some is strictly classified.

Britain's CBW establishments are, on the face of it, small beer compared with those in the United States. Her equivalents to Fort Detrick and Edgewood Arsenal are two laboratories at Porton Down known as the Microbiological Research Establishment and the Chemical Defence Experimental Establishment. Unlike their American counterparts their titles indicate the precise area of their scientific interest; they also differ in that the Ministry of Defence, whose establishments these are, takes trouble to emphasise in its publicity for the laboratories that the work carried on there is of an exclusively defensive nature.

Needless to say, no matter whether this defensive work is in the chemical or biological field, if it is to have any real value it must involve a full understanding of all likely CBW methods of aggression. The ethical dilemma this produces amongst the scientists who work in these establishments today is fundamentally no different from that of Theodor Rosebury working at wartime Camp Detrick, who found the dividing walls between defence and attack to be very thin. But some resolve it. In the words of the director of the Microbiological Research Establishment at Porton, Dr C. E. Gordon Smith,

> As a medical man I am opposed entirely to the offensive use of biological warfare. . . . I think all of us go through a struggle with our conscience when we come here and we would have this struggle more frequently if we were not firmly convinced, and happy in our minds, that what we do is directed towards defending this country and the people.[7]

Much of the work carried on at Porton is published and academic biochemists have acknowledged its high quality. For example, one piece of work reported in 1961 identified erythritol, present in the foetal tissues of cattle, as the substance which stimulates the growth of bacteria responsible for contagious abortion. This obviously valuable livestock research has been described as 'inspired' and 'momentous'. But not all the laboratories' work is made public. Twenty per cent of M.R.E.'s papers remain classified because, according to the director, they contain information which might be of advantage to potential attackers.

Porton, however, contracts some of its work to other institutions, and many people in Britain were surprised to learn in May 1968, as a result of publicity given to a report of a Select Committee on Science and Technology, that projects in several universities were being carried out on behalf of the governmental chemical and biological establishments. None of this work, however, is concerned with weapons systems, but forms a part of normal laboratory studies which are free to be published. Nonetheless, there are close similarities between the manner in which British academic chemists and biologists of the 1960's use the Ministry of Defence establishments, as a golden cow to be milked of its useful funds, and that in which Wernher von Braun used German Army Ordnance in the 1930's.

The fact that work contracted from Porton may be published has, for the most part, meant that there has been no public outcry, as was the case when disclosures were made in America that certain universities there are contributing to restricted CBW research. For several years the University of Pennsylvania was contracted to carry out applied research on weapons systems under U.S. Army and Air Force agreements. Protests from within the University mounted until the spring commencement ceremonies of 1967 when many faculty members flaunted their displeasure in public by appearing in academic dress wearing gas masks. Some time afterwards the University disaffiliated itself from its contracts. But the work continues under its original director, Dr Knut

Krieger, at a laboratory in Philadelphia called the Institute for Cooperative Research. Dr Krieger too maintains that the projects for which he is contracted, known by the code names 'Summit' and 'Spicerack', are defensive in character:

> Basically our tests on both projects is to evaluate what chemical and biological systems would do if used by anyone and how to defend ourselves against them. . . .[8]

But Krieger's thoughts are clearly not limited to defence. He has said:

> . . . I and many of my colleagues on the project, feel that we are opening up a possibility of a kind of warfare that has never been known before and if it really can be accomplished – if we can achieve what we hope to achieve – [will] really be much more humane than ordinary warfare. I know this word 'humane' has been over-used, but it is a fact that we are particularly interested in the so-called incapacitating or non-lethal weapon system. And if we do indeed succeed in creating incapacitating systems and are able to substitute incapacitation for death it appears to me that next to stopping war, this would be an important step forward.[9]

Krieger is not referring here only to tear gases and riot control agents such as Porton's product, CS. He is concerned with a whole new range of products which, once perfected, some at least of the advocates of chemical weapons believe, could mean that warfare might be conducted with quite new concepts. These are the psychochemicals, and they are as yet far from a state of 'perfection'.

The best known is LSD–25 which was first reported in 1943 when the Swiss chemist Albert Hofmann, after absorbing some of the compound during a laboratory experiment, was overcome with a sequence of unusual dreamlike sensations. Since Hofman made his first accidental journey into the quasi-unconscious, the LSD trip had achieved considerable notoriety. Minute quantities of this drug can affect human beings and will bring about unpredictable responses varying from symptoms described as resembling schizophrenia, including delusions of persecution and lack of will to move and communicate, to panic and violent action.

Bizarre experiments have been performed with certain of these psychochemicals. In one, cats exposed to a hallucinogenic gas have turned tail in obvious horror and fright when confronted with a mouse. And in the United States a Congressional committee was greatly impressed by a film of troops exposed to one of these agents. The soldiers were reportedly 'not even conscious of their abnormal condition which was so changed that they were unable to follow simple commands and perform normal tasks with acceptable accuracy'. Brigadier General Rothschild, speaking of incapacitating agents in his book *Tomorrow's Weapons*, states that most of them produce only temporary effects followed by spontaneous, complete recovery. The long-term effects of LSD–25, however, are far from being fully understood. It has been reported widely to be the cause of structural chromosomal alterations and to cause stunted births when fed to animals. One publication in the medical press[10] has described malformations to a child born to a woman who took LSD four times during early pregnancy. The Brigadier, once Assistant Professor of Chemistry at West Point, and a strong advocate of the relative humanity of chemical weapons, goes on,

> Think of the effect of using this type of material covertly on a higher headquarters of a military unit, or overtly on a large organisation! Some military leaders feel that we should not consider using these materials because we do not know exactly what will happen and no clear-cut results can be predicted. But imagine where science would be today if the reaction to trying anything new had been 'Let's not try it until we know what the results will be'.[11]

Whether most scientists would subscribe to the view that covertly to try some new agent on human subjects without knowing the consequences is a humane act is debatable. However, there are many scientists, who have had front-line experience, who can wave their war wounds with the rest and who have spoken up for the humane quality of certain chemical weapons. The geneticist J. B. S. Haldane was one. He pointed out that whereas in the Great War conventional shells killed one man for every three put out of action, mustard gas killed only one man in 40. The scientist's advocacy of the desirability in battle of the early incapacitating agents was not shared by all senior army officers. Haldane writes:

> In 1915 a British chemist proposed to a General who was concerned with such questions that the British should use dichlorethyl sulphide [mustard gas]. 'Does it kill?' asked the General. 'No,' he was told, 'but it will disable enormous numbers of the enemy temporarily.' 'That is no good to us,' said the man of blood, 'we want something that will kill.'[12]

On the other hand, some British military men, contemporaries of Haldane, have agreed with him. Captain Sir Basil Liddell Hart, who was himself a phosgene casualty on the Somme in 1916, was considered so badly disabled as to get the maximum wound gratuity, which was the same as that awarded for the loss of both legs or both arms. Though the effects ruined his military career and turned him into a military historian, he is one of the present-day advocates of the weapon which chemical science has provided which he believes reduces the necessity of killing. He describes the non-lethal gas in Vietnam as 'the most humane weapon ever used in war'.

A British Minister with special responsibility for disarmament has even gone on record as saying that there are circumstances in which the use of non-lethal chemical weapons is not only justifiable but sometimes infinitely preferable to the use of more conventional weapons. Lord Chalfont, himself an ex-soldier, is able to speak from the experience of having personally used tear-gas in many military situations.

But at the present time no incapacitating agents exist which can be used in other than small-scale, above-ground, unenclosed military operations and which can be said to have unfailingly reversible effects; this is true of the frequently used tear-gases just as it is of the psychochemicals which chemists are at this moment trying to perfect. If these agents are to be effective on 100 per cent of the personnel they are intended to incapacitate (and they do not differentiate between soldiers and civilians), large overdoses must inevitably be employed, and these overdoses must in some cases kill.

'Humanity' in war remains a relative term; the possible substitution of 'incapacitation for death', for which Knut Krieger is searching, is still far removed and his attempted goal has been described with some bitterness by other scientists as nothing more than an armament of public relations. The debate concerning the comparative beastliness of chemical and biological weapons has, since the first full-scale chlorine gas attack in battle up to the present time, been charged with emotion. It is not easy to look on new and powerful weapons of war with dispassion, and the emotion raised by the arguments for and against chemical and biological warfare research has tended to cloud the deeper issues involved. Further, it would be a dangerous deception to pretend

that the progress in chemical and biological armaments' manufacture is a greater threat to world peace than the existing uneasy
nuclear arms' balance between the Eastern and Western powers.
But what makes the subject of CBW of such deep concern to the
present is that the scientist and technologist are so refining the
techniques of the armoury as to bring us near to a critical point in
its history.

What many scientists fear is not so much the possibility of an
over-evaluation by over-eager politicians of the humane characteristics of existing so-called incapacitating agents, but that once
they have become an acceptable feature in modern scientific warfare they will throw open the chemist's and biologist's Pandora's
box of weapons. Out of this box might spill, and, given the
precedent, be put to use, some weapons as cruel as any which
dexterous science of the last two decades has invented.

A peak in the sophistication factor of chemical weapons has
long since arrived. According to Brigadier General Rothschild,
the nerve gases which have been developed are so lethal that,
even if new ones could be found, the United States would not
strengthen its arm by owning them. The inhalation of the vapour
from a drop smaller than a pinhead of Sarin (one of the so-called
G-agent nerve gases) can kill. An even smaller droplet of one of
the new less volatile nerve gases (V-agents), deposited on an
exposed area of skin from which it can be absorbed into the
central nervous system, is equally and finally lethal.

In Denver, Colorado, the Rocky Mountain Arsenal is maintained in a continuous stand-by condition for the production of
these nerve gases; moreover, the likely effects of its products have
been publicly demonstrated to the alarm of the citizens of the
State of Utah. One day in March 1968 herds of sheep grazing in
Skull Valley, 20 or 30 miles from one of the United States Army's
chemical and biological weapons testing centres at Dugway,
about 60 miles south-west of Salt Lake City, suddenly and from
no obvious cause began to show alarming symptoms of sickness.
Animals were seen to stagger over their desert pasture as if unable to co-ordinate their limbs, collapse and lie kicking, unable to
rise; within 24 hours most were dead. Losses suffered by sheep
farmers were estimated as 370,000 dollars.

It was later reported that on the previous day the army had
carried out airborne tests with hundreds of gallons of nerve agent

over its proving grounds. The army quickly denied responsibility for the deaths of the sheep. But the case was taken up by Senator Frank E. Moss from Utah who claimed that the results of tests by the United States Department of Agriculture and the Public Health Service proved the military's guilt. No humans were found to have been poisoned, though it has been reported that some workers at Dugway have been affected by nerve gas tests. One theory as to why sheep suffered more than other forms of life in Skull Valley is that they are the only animals known to eat snow, and samples of snow taken from near Dugway were found to be contaminated with the same nerve gases as those used over the test area.

The means for the delivery of nerve agents on distant targets had already been provided by technology; Sarin-filled war-heads are available in the United States for the Honest John rocket and the Sergeant missile. There would be no technological problem in fitting inter-continental ballistic missiles with nerve gas war-heads.

In Britain the Ministry of Defence has patented its own nerve gas processes and delivery systems. Russia, at the most conservative estimate, must at least be keeping pace. Even in 1960 a United States Senate subcommittee on chemical, biological and radiological warfare disarmament accepted a report that over 15 per cent of Soviet munitions in Europe were chemical weapons. And it was in the same year that M. M. Dubinin of the Institute of Physical Chemistry of the Academy of Sciences of the U.S.S.R., speaking of chemical weapons, estimated that '. . . one modern heavy plane with a payload of 10–15 tons may strike a whole area corresponding approximately to that destroyable by a medium or large atomic bomb'.[13]

Thus there exist, even if not yet integrated into one weapon, the components of chemical devices which could be used strategically to annihilate a not inconsiderable proportion of the earth's population. Unlike the atomic bomb, which was born as a result of a calculated political decision and a unified scientific compliance, these weapons have grown from the disparate work of many chemists and biochemists, some of it apparently at first unrelated to warfare, some of it concerned with defence, and some of it knowingly concerned with attack. Unlike the atomic bomb, they have yet to be put to use in war; their strategic value is as uncertain as that of the plutonium weapon which sat in the Alamogordo desert in July 1945; and their effects in terms of the cold measurable statistics of human lives destroyed cannot yet be presented to the world in an easily understandable form, as they were after Hiroshima.

In making predictions for the future use of *their* science in warfare some biologists have laid themselves open to criticisms of wanton sensationalism from some of their colleagues. But the examples of scientific history are not on the side of the critics. It is as well to remember that in 1943 Niels Bohr described the end product of the laboratories of the Manhattan Project, which he was looking at for the first time, as 'what until a few years ago might have been considered a fantastic dream'. After the fulfilment of the dream of the atomic bomb came the fulfilment of what was at one time a piece of even more unrealistic imagination: the hydrogen bomb.

Of the means of waging war biological weapons evoke more than any other sweeping emotions of horror and revulsion. No community has yet had to suffer a large-scale biological attack, but the common experience of the history of all nations has shown what its effects might be. The bubonic and pneumonic plague infections of the 14th century which made up the Black Death were calculated to have killed off 25,000,000 people: a quarter of Europe's population. In the slow spread of London's Great Plague, about 70,000, estimated from the bills of mortality of 1665, perished from a population of 460,000. In more recent times the foot and mouth epidemic, which in 1967 swept through more than 2,000 herds of British cattle, sheep, pigs and goats, demonstrated, to farmers at least, how devastating an economic weapon a virus infection can be: so much so that rumours were heard in some farming communities that the disease had been deliberately planted on the island.

Not only the insidious qualities of biological weapons and their self-propagating nature within human and animal bodies cause alarm. The lethal doses of these agents are small compared even with chemical weapons. Only a few organisms of certain species are required to initiate infections in man. The resistant organism *Coxiella burnetii* which causes the rickettsial disease Q-fever, for example, is considered by CBW experts to be as effective as Sarin in one-millionth of weight by weight quantities of the nerve gas. And biologists with a mathematical bent have calculated that less than 1 lb of botulinus toxin, a non-contagious toxin of *Clostri-dium botulinum,* which was being isolated at wartime Camp Detrick, could, in theory at least, kill the whole of the world's population.

An agent which presents itself for such blanket use in these physically small quantities bring it into line for possible use as a strategic weapon. But the qualities of such a weapon which the biological armourer will demand of his biologist are stringent ones if it is to be truly effective. It must be possible to develop the disease-causing pathogenic organism quickly in suitably large quantities in conditions under which it can rapidly reproduce itself without losing its virulence. It must then be capable of being stored, again without losing its virulence, for sufficiently long periods so that it can be delivered to its distant target. The war-head or other munition which deposits it on its target must be so devised that the organism is not itself destroyed by the pressurised or other mechanical explosive device which distributes it. And once having reached its target human (or animal or food and water supply), then it must there be resistant to attack by natural defence mechanisms of the body, by some effective vaccine or by some other counter-measure. Finally, biological weapons do not respect national or any other artificial frontier; just as the German troops found when the prevailing winds of the Western front blew back their chlorine over their own lines, many of the agents of chemical and biological warfare can be two-edged weapons. Whoever employs a biological agent must himself be able to provide protection from it.

No nation has yet seen fit to announce that its biologists have produced a weapon of this ideality. However, some biologists believe that the ultimate biological armament will come within reach if certain experimental breakthroughs can be brought about.

Already biological engineering has reached such a stage that it is possible to tamper with the genetic material of bacteria and viruses so that their fundamental characteristics can be altered; their infectivity, for example, may be increased or decreased. It is now considered that one day it will be feasible to mask these organisms so that they will not be recognised by the anti-bodies which normally resist them in the human body; when this has been done, a live particle will exist which can multiply in its host, and against which the host will have no pre-existing immunity. What will make the particle more insidious is that no method of early warning, leading to some form of defence, will be available to whomever is made to suffer it since its characteristics will have been masked by the attacker.

A breakthrough of a second kind is considered to be hanging some way over the horizon. When the cells of the human body are damaged in any way, the body counters the damage by producing enzymes which are capable of making good the injury: in other words, the body has its own inbuilt first-aid shop. If it were possible to produce another enzyme, or some other substance which could neutralise this natural reaction, then the body would be robbed of its method of repairing any damage whatever done to it: the first-aid shop would have been abolished. It would not matter whether the injury was caused by bacteria or by the day-to-day wear and tear which the body has to counter from birth. The disease produced by such an agent would be a continuous multiplication of every tiny natural shock that flesh is heir to.

Some scientists believe that the experimental difficulties which must be overcome before these breakthroughs can be brought about make it unlikely that ideal biological weapons will be developed in our time. Yet the manner in which technology applied the principles of physics and chemistry to warfare, when the occasion demanded, has led others to more pessimistic conclusions; some consider that the scope and pace of biological research are now so great that agents of these types could be given every one of the other desirable characteristics required by the biological armourer.

The danger that most powerful biological weapons will sooner or later emerge as a result of work at present underway is real enough since the very purpose of the laboratories which could be responsible for the crucial discoveries could be quite the opposite of life-destroying. For example, knowledge from research now in progress which is aimed at preventing cancerous cells from repairing themselves could be that which leads to an ultimate weapon.

The director of the Microbiological Research Establishment at Porton describes his laboratory's work as being basically preventive medicine. But even if this British establishment were engaged in producing offensive biological weapons (which is officially denied), outwardly it would look no different from what it does today. In fact a laboratory whose purpose is as innocent as that of growing a simple vaccine would appear substantially the same as

322

one committed to preparing virulent strains of micro-organisms.

Not only is the character of the technology of public health fundamentally similar to that of biological warfare, the cost and scale of the technology are the same in both cases. It took the resources of a vast and rich industrialised nation and the combined abilities of the physicists of many Western nations to build the device which raised the first radioactive mushroom cloud over the Alamogordo desert, as it did to produce its thermonuclear successor which finally sank Elugelab Island into the Pacific Ocean. But it is conceivable that the laboratories which will produce devastating strategic biological or chemical weapons will be not sprawling industrial complexes but insignificant clusters of sheds run either by a handful of skilled scientists, or by a few technicians using the knowledge of skilled scientists. It is thus possible that, given the technology to deliver the armament on target, small nations with few resources will before long be able to assemble weapons with the same ultimate annihilatory qualities as those of the hydrogen bomb. There is no reason to suppose that militarily insignificant countries will shy away from choosing the most politically inflammatory toys available from the scientist's Pandora's box. There are several precedents. Mussolini used mustard gas on his barefooted Abyssinian foes, and President Nasser is reported to have used gas on Yemeni villagers. If the worst prophecies of the biologist Cassandras are correct, before the end of the present century every poor dictator will have a chemical agent for his atom bomb, and its biological brother for his thermonuclear weapon.

There is no reason to suppose that small and technologically impoverished nations will have difficulty in recruiting a few scientists of the right quality for the task. Scientific expertise is by now widespread throughout the world. What is not so surely widespread is the tradition of responsibility which has grown up among scientists who, during the first half of the century, have seen the results of their work applied in war: a responsibility born of the belief that recondite knowledge places the scientist in a singular position in society. A little knowledge in the hands of men who are unfamiliar with the growing urgency among scientists to recognise that responsibility could lead to dangerous world-wide consequences. Are there no ways of minimising the hazards in this situation?

16: Tomorrow, and Tomorrow

There is an anecdote that Isaac Newton, who was an elected M.P. for the University of Cambridge, sat out his time in parliament with only one recorded utterance: he asked a steward to shut a window that was causing a draught. If the story is true then his inactivity at Westminster matched up to what was expected of him. It was Robert Hooke, secretary of the Royal Society, of which Newton was later to be president, who in 1663 had implored his fellow natural philosophers to 'improve the knowledge of all natural things . . . by Experiment (not meddling with Divinity, Metaphysics, Moralls, Politicks, Grammar, Rhetoric, or Logick)'.

For two centuries Hooke's advice to the experimentalists was broadly followed, and even through the Victorian age of rationalism, when science had brought the face and form of Europe to a stage of metamorphosis, the moral and political consequences of the scientist's work were not considered by him to be his special concern. The irreversible change of this attitude arrived with the first major war of the 20th century. When it ended in 1918 the influence of applied science on warfare had been so great that it was clear that during the rest of the century the scientific practitioners would never again be able to divorce themselves from the implications of their discoveries as they had so freely done in the past. World War I was fought with chemistry. In World War II it was the physicists who turned their science into weaponry and who now began to recognise their responsibilities. They succeeded only in part in drawing the attention of society to the implications of their work before the worst harm was done and they emerged from that war as intensely valuable pieces of

324

national property; they were cherished according to that value, and in some cases punished according to it when society considered that it had proved that they had turned traitor or heretic.

World War III, if it comes, could conceivably be fought with biology. Biologists have now warned that their science is at that stage which is strictly comparable to that reached by nuclear physics in the few years before World War II. The previous chapter reads as a depressing account of how the technology of this science, now fast rising to the crest of its wave, given a combination of unfortunate circumstances, might lead to – and might lead to the use of – a most powerful device. History can offer only one example of a radical new weapon which has not been put to use in war, and that example – the thermonuclear bomb – is yet to be sustained over the period of a major conflict.

But is this consequence of biology in *its* war as unavoidable as were those of chemistry and physics in theirs? Is it an inevitable irony that the century which Alfred Nobel rang in with offers of rewards to peacemakers from his fortune founded on explosives should end with a death knell rung by the other beneficiaries of his prizes, the scientists? Should we not have known that the promise of total control of our environment, made by both the physical and the life sciences, would not only not be realised and not elevate us to hope and joy, but would in the end reduce us to suspicion and fear? With the knowledge behind us that, in Robert Oppenheimer's words, 'physicists knowingly interfered with history', and of the experience of Hiroshima which was a consequence of this interference, what is our real future?

Any observer of the rate of scientific and technological advances of the past few decades must be myopic if he has not seen in the middle distance some worrying signs of the likely long-term profound effects of these advances. Such is this rate that it is becoming clear that the accumulated benefits are in danger of being outweighed by the social problems simultaneously being created – being created sometimes by these same advances and sometimes by their side effects. The internal combustion engine is one example of a simple invention which as much as any other single piece of technology radically raised the standard of living of the nations able to employ it. Yet within the lifetime of those who saw it introduced, the motor car is beginning to threaten the quality of life in the urban areas where it is most used.

Can we then look with anything other than pessimism on the ultimate consequences of those new advances specifically created, not as benefits but as threats, or advances which it is plain to see can be converted into adjuncts of war? Must we now concede, on the basis of past experience, that all hope of preventing the introduction of yet more weapons by science is dead?

There are signs that all predictions need not be pessimistic. Paradoxically the grave warnings of the biologists can be seen as the brightest ray of hope. It is not simply the fact that these warnings have been given, but that they appear to have been more widely heard and recognised than ever before – and this time not only by a small group of scientists. Loudly, and well ahead of the time that society might feel the effects of their work, some have shouted out and even over-dramatised its implications – on occasions to the embarrassment of some of their colleagues. Learning from the experiences and the dilemmas of the chemists and the physicists, the biologists have acknowledged their responsibilities. Society has been given the chance to prepare itself.

But having made this acknowledgement, how best to ensure action? It was the impact of physics on World War II which finally committed scientists in general to organising themselves so that their moral and political beliefs could be displayed to society. The issues involved now, as then, are too complex and too diffuse to allow science to be heard with one voice. In consequence, whether the many organisations which grew from the post-war years of ferment are as effective as they need be is arguable. But at worse, they function, they are heard, and as a result hope lives on.

The Atomic Scientists of Chicago, the Federation of Atomic Scientists, the Federation of American Scientists and their European counterparts, which stemmed from the disillusioned backwash of the attempts to control atomic power before Hiroshima, have tried to ensure that informed scientists should be able to influence the political decisions which have applied scientific discoveries in the State. Beginning in the 1950's scientists have been increasingly willing to be swept into political life as advisers to governments of the technological nations. They have also, in smaller numbers, entered politics directly. There, their judgments of politics and ethics have been found to be only as

326

illuminating or as fallible as the next man's, but their specialist knowledge has helped to prepare governments in forecasting for the future what the likely influence of science will be.

Less formal bodies of scientists have grouped themselves in attempts to provide platforms from which political solutions can be proposed for the technological problems which their work threatens to thrust down the throats of an unprepared society. When in 1954 fear of the hydrogen bomb reached its height, Bertrand Russell drafted a Manifesto which warned Governments of the catastrophic consequences for mankind of a thermonuclear war. The Manifesto, which became known as the Einstein–Russell Appeal, was signed by Albert Einstein two days before his death. The eleven co-signees included nine Nobel prizewinners, most of them physicists and therefore intimately aware of the issues at stake. This was the starting point for the 'Pugwash' Movement of scientists financed by the industrialist, Cyrus Eaton, which held its first meeting at the fishing village of Pugwash in Nova Scotia. Its conferences have always been attended by scientists from the upper stratum of their profession so that their attempts to formulate solutions to the problems facing mankind can, in theory at least, be passed on to comparably influential levels of government.

Other groups have formed themselves as a result of deep moral concern and the wish to see some limits set to the way in which science allows itself to be employed in the war machine. The British Society for Social Responsibility in Science, which held its inaugural meeting on the Royal Society premises in April, 1969, had its grassroots in a chemical and biological warfare conference held in London in the previous year. It specifically did not restrict its membership to scientists from the top drawer and sought to concern itself with a wide range of implications and consequences of scientific development to which it hopes to alert both science and the public.

Many societies such as these – and there were similar ones in Britain formed before the Second World War – after the initial head of steam has been given off, run out of momentum and membership and remain only as footnotes of idealist hopes in history books. The reasons for this are plain. It is easy to define the broad social problems presented by scientific advances but difficult to provide workable specific solutions. Conceived in the

United States twenty years before the British society of the same name, the Society for Social Responsibility in Science (with which again Einstein associated himself), set for itself the tasks,

> . . . to foster throughout the world a . . . tradition of personal moral responsibility for the consequences for humanity of professional activity, with emphasis on constructive alternatives to militarism; to embody in this tradition the principle that the individual must abstain from destructive work and devote himself to constructive work, according to his own moral judgment; to ascertain . . . the boundary between constructive and destructive work, to serve as a guide for individual and group decisions and action. . . .[1]

But it is because this boundary can seldom be seen with clarity and can be raised or lowered according to the height of the individual's moral judgment that the unified cry of the conscience of science cannot be heard and the scientists' dilemma still exists in spite of themselves. There is still widespread disagreement as to whether and how science should be involved in decisions relating to warfare. The major conflict of the post-war years had been between disarmament and deterrence. Hans Bethe, the physicist who worked on the hydrogen bomb hoping to prove that it could never be built, wrote in 1958:

> In order to fulfill this function of contributing to the decision-making process, scientists (at least some of them) . . . must be willing to work on weapons. They must do this also because our present struggle is (fortunately) not carried on in actual warfare which has become an absurdity, but in technical development for a potential war which nobody expects to come. The scientists must preserve the precarious balance of armament which would make it disastrous for either side to start a war. Only then can we argue for and embark on more constructive ventures like disarmament and international co-operation which may eventually lead to a more definite peace.[2]

Many would strongly agree with him; but of course equally strongly many would not. In 1957, 18 distinguished German scientists signed a declaration which urged the Federal Republic of Germany to renounce both the development and the possession of nuclear weapons of any kind. Work on biological weapons produces the same conflict. Professor Maynard Smith, the British geneticist broadcasts that 'people who work at Porton on germ warfare ought to be bloody well ashamed of themselves'[3] for doing research which might ultimately lead to a means of destruction of our species. However, Professor E. B. Chain, who shared a Nobel Prize for his work on penicillin, writes to a national newspaper:

> As far as the excellent team of scientists at the Microbiological Establishment at Porton, by international consensus of opinion one of the finest in the world, is concerned, do we not have to consider ourselves lucky to have in this country such a highly competent assembly of experts who are able to assess the effectiveness of biological warfare in attack and to develop defensive methods should a ruthless enemy, as he well might, use these techniques for attack against our people?[4]

The issues involved are not clear-cut and never have been. One day in 1935 Robert Watson-Watt asked his assistant, Arnold Wilkins, to calculate how much radio frequency power would be required to heat up a volume of liquid, equal to the amount of blood in a man's body, through a temperature of 100 degrees Fahrenheit at two miles' distance. Yet how many gentlemen of England would now feel inclined to say that Watson-Watt ought to be bloody well ashamed of himself for having considered this experimental possibility in his work which preceded the invention of Britain's most significant piece of World War II defence, radar? And how many Americans, deeply suspicious of work in progress in biological back-rooms, would see fit to condemn the plant physiologist Professor Arthur Galston of Yale University who, in his Ph.D. thesis of 25 years ago, presented a technique for inducing flowers in soya bean, the present applications of which are to produce large increases in the bean crop yield, and also as a defoliant in Vietnam.

Inevitably, as scientists have acknowledged their responsibilities the cut and thrust among them has become more vicious. Were it possible to devise some rigorous ethical code of conduct for scientists then the conflict might resolve itself. For centuries physicians have been able to refer to their Hippocratic oath to perform their art 'never with a view to injury and wrong-doing'. But with the advances of biology of the last decade presenting them with means of transplant surgery to save life, a rigorous definition of death of the donor of the human organ becomes vital. It is now no longer sufficient to accept the stopping of the heartbeat for this definition. As medical research progresses the sharp moral line which governed behaviour becomes less easily defined. Doctors too are now experiencing the dilemma of scientists who have never had delineated for them any boundary to govern their ethics. And, as in recent years genetic and biochemical advances give rise to the possibility that one day scientists may be responsible for controlling both human and psychological characteristics in the laboratory, so the difficulties of defining such a guideline become more perplexing. It could be easy to condemn such men as Robert Oppenheimer, Edward Teller, and Hans Bethe, who own patents on the nuclear and thermonuclear weapons and who have come to terms with their consciences. It could be easy to absolve those men at present working in cancer research laboratories or in breweries whose work might lead to the ultimate biological weapon and who have never had a problem of conscience. It is not possible to point to the line where condemnation should begin and absolution end. Yet in spite of the obvious problems, at the 17th Pugwash Conference in 1967 the Committee on Special Responsibility of Scientists recommended that a study group should be set up to attempt to formulate ethical guidelines for scientists.

Traditionally science has been a search for the truth, and it is this maxim on which the worldwide brotherhood of science was built. Many factors have contributed to its breakdown. Chief among these has been the application of science: its application to benefit mankind and to threaten it. Even if he wanted to be completely detached from the surrounding world (which most scientists do not) and not care whether his work is applied in society, the scientist cannot remain of choice in a cocoon. Willy-nilly his work is applied and affects society and this in turn perturbs the original concept of science. There is an uncertainty relationship in existence between science and the community; the position and momentum of the former at any given time cannot be clearly defined. Nationalism and chauvinism too have raised their ugly heads in science (as is evident when the credit for some new discovery comes in dispute). The State, which looks on itself as the legitimate guardian of national resources, has welcomed with open arms as many of these applied offerings as it can consider useful. And since every State, and there are no exceptions, allows that war is a valid instrument of national policy, science finds itself up to the hilt in warfare, and itself involved in internal conflict.

Science can never return to its former state of innocence. It can, however, take steps towards re-establishing some form of the brotherhood which once existed without discarding its responsibilities, and by so doing confer enormous benefits on mankind. The days are long since past when a scientist working in one particular field could reasonably hope to keep his eye on the main progress in research in all other fields. The world today houses 5,000,000 scientists and engineers and this number is expected to increase to 25,000,000 by the year 2,000 A.D. The problems of monitoring the vast output which such a population of workers produces are truly enormous. But provided this work is put into the public domain so that it can be monitored by scientists and non-scientists alike, so that it can be understood and its implications can be examined, then the dangers from its misuse will be greatly diminished.

When the spectrum of science was narrow, when it considered itself to be concerned only with a search for truth and not with its applications, the brotherhood was sustained by an untroubled, unrestricted publication of data. It is conceivable that the most beneficial aspects of the brotherhood could be retrieved if this state of affairs were reimposed and if some attempt could be made to limit or even abolish the amount of secret research which is carried out by scientists on behalf of governments throughout the world. Many of the scientists who have organised themselves into groups attempting to influence political policy have expressed this view. Some of those biologists concerned with the Pugwash movement, for example, believe that control over biological weapon development can now only be achieved by international collaboration and that this can only be approached if open research and publication are reinstituted. In 1964 a group of scientists of the movement began to study and test the feasibility of a system of voluntary inspection of laboratories engaged in microbiological research and manufacture. With cooperation from the countries concerned, institutions in Austria, Czechoslovakia, Denmark and Sweden were visited and investigated, the long-term aim being to expand the scheme on a world-wide basis. If all laboratories could be thrown open to inspection of this kind and a viable open research agreement reached, and if the internationalism amongst scientists were to become so powerful that infringements of agreements could easily be detected, then the fear of some massive

331

mistake leading to a biological holocaust could be removed.

Many British scientists believe that a start could be made at the Porton laboratories and that its defensive work should be laid bare for public investigation. One suggestion that has been made is that Porton could be declared declassified by the British Government for an initial trial period. Such a gesture would be one of world leadership, would put public opinion to a realistic test, and would have the advantage of not breaking up the skilled scientific teams which have been built up by the Ministry of Defence.

There have even been suggestions that conditions should be created where a scientist, under certain circumstances, could be permitted to take unilateral action when his work conflicts with the dictates of his conscience. Thus some biologists that where a country has promised, by ratifying an agreement to prohibit biological weapons, scientists of that country should have the right to disclose any evidence they have that the agreement is being broken. A scientist who did this would therefore be practising open espionage. However, as history has shown, there are most serious dangers involved when one man takes it on himself to become the arbiter of what is or what is not to the advantage of the society of which he is a part, and acts accordingly; the cases of Allan Nunn May and Klaus Fuchs exemplified only a few of the pitfalls.

But *all* suggestions to open up secret war weapons to the eyes of potential enemies have intrinsic and obvious dangers, though the reasons prompting these suggestions are neither naïve nor new. Niels Bohr and Robert Oppenheimer clearly saw in 1945 the mistrust and the fear which the owners of the first atomic weapon would generate in Russia, just as they saw that Russia would sooner or later build its own nuclear weapons and succeed in balancing fear with fear. If the mistrust resulting from secrecy which drives biologists to persist in weapon work could be removed, the ultimate biological weapon need never be created to join its thermonuclear counterpart.

This ideality, this scientific heaven, might never arrive. But confident and invaluable steps have been taken towards more outrageous ideals; the elixir of life was never distilled in a chemist's laboratory but gerontologists are far from being despondent about biological discoveries which will extend the useful

human life-span; and the philosopher's stone has never material-
ised, but physicists have nonetheless found methods of transmut-
ing one element into another. Equally unexpected and beneficial
results might push through cracks made by an assault on the
apparently insuperable barriers in the path of completely open
scientific research. The stakes are so high that the ideal must be
worth struggling for.

If the brotherhood is ever built up again, then the precept of
science for science's sake will have been left far behind. The
conscience of science has only developed with the realisation that
science in the end cannot exist for itself, but must be for the sake
of society. The scientist is the first in line as custodian of society's
ability to handle the forces of nature. Danger arises only when the
scientist abandons this pragmatic base and fails to keep society
and its governments informed of the understanding he has gained
from his specialised knowledge, or when the scientist wilfully
considers himself to be a different breed from the rest of mankind,
and attempts to apply his work without first having consulted
society.

But the responsibility does not end with the scientist. The
French revolutionary, Fouquier-Tinville, said at Lavoisier's trial,
'The Republic has no need of scientists', and Lavoisier's head
rolled; this tragic attitude still manifests itself even now, two
centuries later. C. P. Snow's picture of 20th-century society
split into two cultures has been much maligned; but the schism
exists for all that. Not to recognise it might be to put our heads
on the guillotine, just as surely as was, in the end, Fouquier-
Tinville's. If, however, society outside science is not prepared to
lean forward and listen to what science has to say, just as readily
in the past it has stretched out its hand and taken what science
has had to offer: if it refuses to prepare itself, or to let itself
be educated so that it can begin to comprehend the scientist's
specialised information, then the struggles of science with its
conscience will have been for nothing. In the words of Albert
Einstein, it is science which has brought forth the danger, but
ultimately the real problem is in the minds and hearts of men.

Acknowledgments and Some Notes on Sources

My thanks are due to many scientists and non-scientists who have given me their valuable time in discussions. I also wish to thank: Professor S. A. Goudsmit for copies of letters and papers collected by him on the Alsos mission; Professor R. V. Jones for copies of letters; Professor O. R. Frisch and Sir Rudolf Peierls for permission to quote from the Frisch-Peierls Memorandum; the B.B.C. for permission to quote from *Einstein, The Building of the Bomb, The Debate about Chemical Warfare, The Shape of War to Come,* and *Too Near the Sun*; the Librarian of the Royal Institution for Sir James Dewar's Notes for the Managers of the Royal Institution; William Heinemann for the writings of Alfred Nobel from *The Life of Alfred Nobel,* by H. Schück and R. Sohlman; Deutsche Verlags Anstalt for extracts from *Memoiren* by Bertha von Suttner; Methuen for letters of Albert Einstein from *Einstein on Peace,* edited by Otto Nathan and Heinz Norden; Cambridge University Press for letters from *Rutherford* by A. S. Eve; Oxford University Press for extracts from *Atomic Quest,* by Arthur Compton; Andre Deutsch for extracts from *Now It Can Be Told,* by Leslie R. Groves; Martin Secker and Warburg for letters from *Alsos,* by S. A. Goudsmit; Macmillan for extracts from *Britain and Atomic Energy,* by Margaret Gowing; the Editor for extracts from *Nature*; Cassell for extracts from *The Second World War,* by Winston S. Churchill; Harper and Row for extracts from *On Active Service in Peace and War,* by Henry L. Stimson and McGeorge Bundy, and for extracts from *The Journals of David E. Lilienthal*; the Queen's Printer, Ottawa, Canada for extracts from The Report of the Royal Commission on espionage; and the Editor for extracts from the *Bulletin of the Atomic Scientists.*

Notes and Sources

1. SCANDAL

1. *The Times,* 30th January 1894.
2. *The Times,* 9th February 1894.
3. *The Times,* 15th February 1894.
4. *The Times,* 19th July 1894.

2. THE PRIZE

1. Schück, H., and Sohlman, R., *The Life of Alfred Nobel* (Heinemann, London, 1929), p. 174.
2. Bergengren, E., *Alfred Nobel* (Nelson, London, 1962), p. 65.
3. Schück, H., and Sohlman, R., op. cit., p. 66.
4. Ibid., p. 54.
5. von Suttner, B., *Memoiren* (Deutsche Verlags Anstalt, 1909), p. 131.
6. Ibid., p. 134.
7. Bergengren, E., op. cit., p. 190.
8. von Suttner, B., op. cit., p. 183.
9. Ibid., p. 237.
10. Ibid., p. 271.
11. Schück, H., and Sohlman, R., op. cit., p. 208.
12. von Suttner, B., op. cit., p. 272.
13. *Nobel Foundation Calendar* (Nobel Foundation, Stockholm, 1967).
14. Nathan, O., and Norden, H. (Editors), *Einstein on Peace* (Methuen, London, 1963), p. 355.

3. THE YOUNG GENIUS

1. Richter, W., *Bismarck* (Macdonald, London, 1964), p. 207.
2. *Einstein,* B.B.C.-TV production, 27th April 1965.
3. Ibid.
4. Frank, P., *Einstein* (Cape, London, 1948), p. 141.
5. Nicolai, G. F., *The Biology of War* (Dent, London, 1919), p. 2.

4. THE GREAT WAR

1. Report of a Special Meeting of the Managers of the Royal Institution, 16th May 1916; Royal Institution of Great Britain, London.
2. Willstätter, R., 'Fritz Haber zum sechzigsten Geburtstag', *Die Naturwissenschaften*, 14th December 1928, p. 1057.
3. Ibid., p. 1058.
4. Eve, A. S., *Rutherford* (Cambridge University Press, 1939), p. 245.
5. Rutherford, E., 'H. G. J. Moseley, 1887–1915', *Proceedings of the Royal Society*, A. 1917, p. xxii.
6. 'Henry Gwyn Jeffreys Moseley', *Nature*, 9th September 1915, p. 33.
7. Report of a Special Meeting of the Managers of the Royal Institution, 16th May 1916.
8. Nobel Conference, 1903.
9. Curie, E., *Madame Curie* (Heinemann, London, 1938), p. 349.
10. Eve, A. S., op. cit., p. 264.

5. THE PHYSICIST MATURES

1. Einstein, A., *The World As I See It* (The Bodley Head, London, 1935), p. 4.
2. Nathan, O., and Norden, H., op. cit., p. 54.
3. Ibid., p. 74.
4. Frank, P., op. cit., p. 189.
5. Nathan, O., and Norden, H., op. cit., p. 95.
6. Einstein, A., op. cit., p. 4.
7. *Einstein*, B.B.C.-TV production, 27th April 1965.
8. From a document in the possession of S. A. Goudsmit.
9. Einstein, A., op. cit., p. 90.
10. Nathan, O., and Norden, H., op. cit., p. 229.
11. Ibid., p. 236.
12. Ibid., p. 257.

7. IN GERMANY

1. Groves, L. R., *Now It Can Be Told* (André Deutsch, London, 1963), p. 247.
2. Interview with journalist, October 1950.
3. Klee, E., and Merk, O., *The Birth of the Missile* (Harrap, London, 1965), p. 89.
4. From a document in the possession of S. A. Goudsmit.
5. Ibid.
6. Groves, op. cit., p. 244.
7. Interview with the author, November 1964.
8. Ibid.
9. Dornberger, W., *V2* (Hurst and Blackett, London, 1954), p. 29.
10. From a document in the possession of S. A. Goudsmit.
11. Letter to author, May 1968.
12. Ibid.

13. Groves, L. R., op. cit., p. 248.
14. Ibid., p. 333.
15. Ibid., pp. 334–5.
16. Ibid., p. 336.
17. Heisenberg, W., 'Research in Germany on the Technical Applications of Atomic Energy', *Nature*, 16th August 1947, p. 211.
18. Ibid.
19. Groves, L. R., op. cit., p. 335.
20. Ibid.

8. IN BRITAIN

1. Churchill, W. S., *The Second World War* (Cassell, London, 1949), Vol. 2, p. 338.
2. The Cherwell Papers.
3. Snow, C. P., *Science and Government* (Oxford University Press, London, 1961), p. 23.
4. Ibid., p. 33.
5. Birkenhead, The Earl of, *The Prof in Two Worlds* (Collins, London, 1961), p. 195.
6. Clark, R. W., *Tizard* (Methuen, London, 1965), p. 306.

9. WARNING NOISES

1. Gowing, M., *Britain and Atomic Energy (1939–1945)* (Macmillan, London, 1964), p. 394.
2. Ibid., p. 398.

10. IN AMERICA

1. *In the Matter of J. Robert Oppenheimer,* Transcript of Hearing before Personnel Security Board, U.S. Atomic Energy Commission, Washington, 1954, p. 393 (referred to subsequently as *In the Matter of J. Robert Oppenheimer, Transcript*).
2. Compton, A. H., *Atomic Quest* (Oxford University Press, London, 1956), p. 64.
3. Ibid., p. 151.
4. Ibid., p. 207.
5. *The Building of the Bomb*, B.B.C.-TV production, 2nd March 1965.
6. *In the Matter of J. Robert Oppenheimer, Transcript*, p. 8.
7. *The Building of the Bomb,* B.B.C.-TV production, 2nd March 1965.
8. Ibid.
9. Oppenheimer, J. R., 'Three Lectures on Niels Bohr and his Times', *Pegram Lectures,* Brookhaven National Laboratory, 1963.
10. Letter to author, July 1968.

11. THE COURSE OF HUMAN HISTORY

1. Churchill, W. S., op. cit., Vol. 2, p. 340.

2. Jones, R. V., 'Winston Leonard Spencer Churchill', *Biographical Memoirs of Fellows of the Royal Society*, November 1966, p. 93.
3. Bohr, N., *Open Letter to the United Nations*, 9th July 1950.
4. Stimson, H. L., 'The Decision to Use the A-Bomb', *Harper's Magazine*, February 1947.
5. Ibid.
6. *The Building of the Bomb*, B.B.C.-TV production, 2nd March 1965.
7. Frank, J. *et al.*, 'A Report to the Secretary of War', in *The Atomic Age* (Basic Books, New York, 1962), p. 24.
8. Stimson, H. L., op. cit.
9. Szilard, L., 'A Petition to the President of the United States', in *The Atomic Age* (Basic Books, New York, 1962), p. 28.
10. Teller, E., *The Legacy of Hiroshima* (Macmillan, London, 1962), p. 13.
11. Oppenheimer, J. R., interview with author, October 1964.
12. Teller, E., interview with author, October 1964.
13. Compton, A. H., op. cit., p. 242.
14. Ibid., p. 247.
15. Truman, H. S., *Year of Decision, 1945* (Hodder and Stoughton, London, 1955). p. 350.
16. Churchill, W. S., op. cit., Vol. 6, p. 553.
17. Compton, A. H., op. cit., p. 258.
18. *The Building of the Bomb*, B.B.C.-TV production, 2nd March 1965.
19. Ibid.
20. Ibid.
21. Ibid.
22. Ibid.

12. TREACHERY

1. This and subsequent quotations referring to Allan Nunn May and the Gouzenko affair are extracted from the *Report of the Royal Commission* on espionage, Canadian Government, 27th June 1946. Quotations from May's hearings and trial are extracted from contemporary newspaper reports.
2. Quotations from Klaus Fuchs's trial and from his statement are taken from *Soviet Atomic Espionage*, U.S. Government Printing Office, Washington, 1951, and from contemporary newspaper reports.
3. Monod, J., Letter to the Editor, *Bulletin of the Atomic Scientists*, October 1953.
4. Morrison, P., letter to author, June 1968.
5. Langer, E., 'The Case of Morton Sobell', *Science*, 23rd September 1966.

13. WE BUILT ONE FRANKENSTEIN

1. *Too Near the Sun*, B.B.C.-TV production, 9th June 1966.

2. Ibid.
3. Allison, S. K., 'The State of Physics', *Bulletin of the Atomic Scientists,* January 1950.
4. Baruch, B. M., *The Public Years* (Odhams Press, London, 1961), p. 334.
5. *In the Matter of J. Robert Oppenheimer, Transcript,* p. 273.
6. Ibid., p. 242.
7. Lilienthal, D. E., *The Journals of David E. Lilienthal* (Harper and Row, New York, 1964), p. 581.
8. *In the Matter of J. Robert Oppenheimer, Transcript,* p. 79.
9. Ibid.
10. Bethe, H. A., 'The Hydrogen Bomb, II', *Scientific American,* April 1950.
11. *In the Matter of J. Robert Oppenheimer, Transcript,* p. 305.
12. Ibid., p. 330.

14. THE EXTRAORDINARY PERSONAGE

1. Lilienthal, D. E., op. cit., p. 69.
2. *In the Matter of J. Robert Oppenheimer, Transcript,* p. 16.
3. Ibid., p. 20.
4. Chevalier, H., *The Story of a Friendship* (Braziller, New York, 1965), p. 107.
5. *In the Matter of J. Robert Oppenheimer, Transcript,* p. 130.
6. Ibid., p. 871.
7. Ibid., p. 873.
8. Ibid., p. 149.
9. Ibid., p. 70.
10. Ibid., p. 710.
11. Ibid., p. 726.
12. Ibid.
13. Ibid., p. 262.
14. Ibid., p. 280.
15. Ibid., p. 491.
16. *In the Matter of J. Robert Oppenheimer,* Texts of Principal Documents and Letters, U.S. Atomic Energy Commission, Washington, 1954, p. 1.
17. Ibid., p. 21.
18. Ibid.
19. Ibid., p. 22.
20. Ibid., p. 23.
21. Ibid., p. 17.

15. THE SHAPE OF WAR TO COME

1. *In the Matter of J. Robert Oppenheimer, Transcript,* p. 803.
2. 'Chemical-Biological-Radiological Warfare and Its Disarmament Aspects', *U.S. Senate Subcommittee Study,* Washington, 29th August 1960.

3. Lilienthal, D. E., op. cit., p. 199.
4. 'The Shape of War to Come', *Horizon,* B.B.C.-TV production, 25th April 1967.
5. Fieser, L. F., *The Scientific Method* (Reinhold, New York, 1964), p. 133.
6. Rosebury, T., 'Medical Ethics and Biological Warfare', *Perspectives in Biology and Medicine,* Summer 1963, p. 514.
7. 'The Shape of War to Come', *Horizon,* B.B.C.-TV production, 24th April 1967.
8. *The Debate About Chemical Warfare,* B.B.C. Third Programme production, 11th October 1967.
9. Ibid.
10. Zellweger, H., 'Is Lysergic-Acid Dethylamide a Teratogen?', *The Lancet,* 18th November 1967.
11. Rothschild, J. H., *Tomorrow's Weapons* (McGraw-Hill, New York, 1964), p. 43.
12. Haldane, J. B. S., *Callinicus – A Defence of Chemical Warfare* (Kegan Paul, Trench, Trubner, London, 1925), p. 48.
13. Dubinin, M. M., 'Potentialities of Chemical Warfare', *Bulletin of the Atomic Scientists,* June 1960, p. 251.

16. TOMORROW, AND TOMORROW

1. Bry, I., and Doe, J., 'War and Men of Science', *Science,* 11th November 1955, p. 913.
2. Bethe, H. A., Book Review, *Bulletin of the Atomic Scientists,* December 1958, p. 428.
3. 'Secret Science', *The Listener,* 9th November 1967, p. 606.
4. Letter to *The Observer,* 2nd June 1968.

Select Bibliography
_____

Allison, S. K., 'The State of Physics', *Bulletin of the Atomic Scientists*, January 1950.
Baruch, B. M., *The Public Years* (Odhams Press, London, 1961).
Bennett, G., *Coronel and the Falklands* (Batsford, London, 1962).
Bergengren, E., *Alfred Nobel* (Thomas Nelson, London, 1962).
Bethe, H. A., 'Oppenheimer', *Science*, 3rd March 1967.
Bethe, H. A., 'The Hydrogen Bomb, II', *Scientific American*, April 1950.
'Biological and Chemical Warfare. An International Symposium', *Bulletin of the Atomic Scientists*, June 1960.
Birkenhead, The Earl of, *The Prof in Two Worlds* (Collins, London, 1961).
Blackett, P. M. S., *Studies of War* (Oliver & Boyd, Edinburgh and London, 1962).
Born, M., 'What is left to hope for?', *Bulletin of the Atomic Scientists*, April 1964.
von Braun, W., and Ordway, F. I., *History of Rocketry and Space Travel* (Nelson, London, 1967).
Brod, M., *Franz Kafka* (Schocken Books, New York, 1964).
Bry, I., and Doe, J., 'War and Men of Science', *Science*, 11th November 1955.
Chemical-Biological-Radiological Warfare and its Disarmament Aspects (U.S. Government Printing Office, Washingon, 1960).
Chevalier, H., *The Story of a Friendship* (George Braziller, New York, 1965).
Churchill, W. S., *The Second World War* (Cassell, London, 1949).
Clark, R. W., *Tizard* (Methuen, London, 1965).
Clarke, R., 'Biological Warfare', *Science Journal*, November 1966.
Compton, A. H., *Atomic Quest* (Oxford University Press, London, 1956).
Curie, E., *Madame Curie* (William Heinemann, London, 1938).
Curie, M., *Pierre Curie* (Payot, Paris, 1924).
Dornberger, W., *V*2 (Hurst & Blackett, London, 1954).

341

Select Bibliography

Einstein, A., *The World As I See It* (The Bodley Head, London, 1935).
Einstein, B.B.C.-TV production, 27th April 1965.
Eve, A. S., *Rutherford* (Cambridge University Press, 1939).
Fieser, L. F., *The Scientific Method* (Reinhold, New York, 1964).
Franck, P., *Einstein* (Jonathan Cape, London, 1948).
Fuchs, E., *Christ in Catastrophe* (Pendle Hill, Wallingford, Pa., 1949).
Gardner, B., *Alleuby* (Cassell, London, 1965).
Gilpin, R., *American Scientists and Nuclear Weapons Policy* (Princeton University Press, 1962).
Goran, M., *The Story of Fritz Haber* (University of Oklahoma Press, 1967).
Goudsmit, S. A., *Alsos* (Sigma Books, London, 1947).
Gowing, M., *Britain and Atomic Energy* (1939–1945) (Macmillan, London, 1964).
Grodzins, M., and Rabinowitch, E. (Editors), *The Atomic Age* (Basic Books, New York, 1963).
Groves, L. R., *Now It Can Be Told* (André Deutsch, London, 1963).
Haldane, J. B. S., *Callinicus – A Defence of Chemical Warfare* (Kegan Paul, Trench, Trubner & Co., London, 1925).
Heisenberg, W., 'Research in Germany on the Technical Application of Atomic Energy', *Nature*, 16th August 1947.
Hewlett, R. G., and Anderson, O. E., *The New World* (Pennsylvania State University Press, 1962).
History of the Atomic Bomb, Radiotelevisione Italiana TV production, 1963.
In the Matter of J. Robert Oppenheimer, Transcript of Hearing before Personnel Security Board, Texts of Principal Documents and Letters, U.S. Atomic Energy Commission, Washington, 1954.
Jones, R. V., 'Infrared Detection in British Air Defence, 1935–38', *Infrared Physics,* July 1961.
Jones, R. V., 'Scientists and Statesmen: The Example of Sir Henry Tizard', *Minerva*, Winter 1966.
Jones, R. V., 'Sir Henry Tizard', *Nature*, 6th March 1965.
Jones, R. V., 'Winston Leonard Spencer Churchill', *Biographical Memoirs of Fellows of The Royal Society*, November 1966.
Jungk, R., *Brighter Than a Thousand Suns* (Penguin Books, Harmondsworth, 1964).
Klee, E., and Merk, O., *The Birth of the Missile* (George G. Harrap, London, 1965).
Knebel, F., and Bailey, C. W., 'The Fight Over the A-Bomb', *Look*, 13th August 1963.
Langer, E., 'Chemical and Biological Warfare, I and II', *Science*, 13th and 20th January 1967.
Langer, E., 'The Case of Morton Sobell', *Science*, 23rd September 1966.
Lilienthal, D. E., *The Journals of David E. Lilienthal* (Harper & Row, New York, 1964).
Marwick, C., 'Death in Skull Valley', *New Scientist*, 25th April 1968.
Meinertzhagen, B., *Middle East Diary* (Cressor Press, London, 1959).

Miles, F. D., *A History of Research in the Nobel Division of I.C.I.* (Imperial Chemical Industries, 1955).

Monod, J., Letter to the Editor, *Bulletin of the Atomic Scientists,* October 1953.

Moorehead, A., *The Traitors* (Hamish Hamilton, London, 1952).

Nathan, O., and Norden, H. (Editors), *Einstein on Peace* (Methuen, London 1963).

Nicolai, G. F., *The Biology of War* (Dent, London, 1919).

Nobel Foundation Calendar (Nobel Foundation, Stockholm, 1967).

Oppenheimer, J. R., 'Three Lectures on Niels Bohr and his Times', *Pegram Lectures,* Brookhaven National Laboratory, 1963.

Passant, E. J., *Germany, 1815–1945* (Cambridge University Press, London, 1959).

Planck, M., *A Scientific Autobiography* (Williams & Norgate, London, 1959).

Reich, P., and Sidel, V. W., 'Current Concepts: Napalm', *New England Journal of Medicine,* 13th July 1967.

Report of the Royal Commission on Espionage, Canadian Government, 27th June 1946.

Richter, W., *Bismarck* (Macdonald, London, 1964).

Ridenouer, L. N., 'The Hydrogen Bomb', *Scientific American,* April 1950.

Robinson, J. P., 'Chemical Warfare', *Science Journal,* April 1967.

Rosebury, T., 'Medical Ethics and Biological Warfare', *Perspectives in Biology and Medicine,* Summer 1963.

Rotblat, J., *Pugwash,* Czechoslovak Academy of Sciences, 1967.

Rotblat, J., *Science and World Affairs* (Dawsons of Pall Mall, London, 1962).

Rothschild, J. H., *Tomorrow's Weapons* (McGraw-Hill, New York, 1964).

Rozental, S. (Editor), *Niels Bohr,* North-Holland Publishing Co., Amsterdam, 1967.

Rutherford, E., 'H. G. J. Moseley, 1887–1915', *Proceedings of the Royal Society,* A 93, October 1917.

Schilpp, P. A. (Editor), *Albert Einstein* (Harper, New York, 1959).

Schück, H. and Sohlman, R., *The Life of Alfred Nobel* (Heinemann, London, 1929).

'Science and Human Welfare', *Science,* 8th July 1960.

Shils, E., 'Leo Szilard', *Encounter,* December 1964.

Sidel, V. W., and Goldwyn, R. M., 'Chemical and Biological Weapons – A Primer', *New England Journal of Medicine,* 6th January 1966.

Smith, A. K., *A Peril and a Hope* (University of Chicago Press, 1965).

Smith, A. K., 'Behind the Decision to Use the Atomic Bomb', *Bulletin of the Atomic Scientists,* October 1958.

Smith, C. E. G., 'In Defence of Defensive Warfare', *Nature,* 3rd August 1968.

Smith, C. E. G., 'Microbiological Research at Porton', *Nature,* 22nd June 1968.

Smith, C. E. G., 'The Microbiological Research Establishment at Porton', *Chemistry and Industry*, 4th March 1967.

Smyth, H. D., *Atomic Energy for Military Purposes* (U.S. Government Printing Office, Washingon, 1945).

Snow, C. P., *Science and Government* (Oxford University Press, London, 1961).

Soviet Atomic Espionage (U.S. Government Printing Office, Washington, 1951).

Stewart, I., *Organizing Scientific Research for War* (Little, Brown, Boston, 1948).

Stimson, H. L., 'The Decision to Use the A-Bomb', *Harper's Magazine*, February 1947.

Strauss, L. L., *Men and Decisions* (Macmillan, London, 1963).

von Suttner, B., *Memoiren* (Deutsche Verlags Anstalt, 1909).

Teller, E., *The Legacy of Hiroshima* (Macmillan, London, 1962).

The Building of the Bomb, B.B.C.-TV production, 2nd March 1965.

The Debate About Chemical Warfare, B.B.C. Third Programme production, 11th October 1967.

The Decision to Drop the Bomb, N.B.C.-TV production, 6th January 1965.

'The Oppenheimer Case', *Bulletin of the Atomic Scientists*, May 1954.

'The Shape of War to Come', *Horizon*, B.B.C.-TV production, 25th April 1967.

'The War of the Boffins', *Horizon*, B.B.C.-TV production, 12th September 1967.

Tizard, H., *The Passing World,* Presidential Address to the British Association, 1948.

Too Near the Sun, B.B.C.-TV production, 9th June 1966.

Trotter, R. (Editor), *The History of Nobel's Explosives Company Limited* (Imperial Chemical Industries, 1938).

Truman, H. S., *Year of Decisions,* 1945 (Hodder & Stoughton, London, 1955).

'Vietnam: the Horrors Multiply', *New Statesman*, 26th March 1965.

Wada, S., et al., 'Mustard Gas as a Cause of Respiratory Neoplasia in Man', *The Lancet,* 1st June 1968.

West, R., *The Meaning of Treason* (Penguin Books, Harmondsworth, 1965).

Willstätter, R., 'Fritz Haber zum sechzigsten Geburtstag', *Die Naturwissenschaften*, 14th December 1928.

Zellweger, H., 'Is Lysergic-Acid Dethylamide a Teratogen?', *The Lancet,* 18th November 1967.

Index

Index

Index